SCIENCE TRANSFORMED?

Science Transformed?

DEBATING CLAIMS OF AN EPOCHAL BREAK

EDITED BY

Alfred Nordmann, Hans Radder,
and Gregor Schiemann

UNIVERSITY OF PITTSBURGH PRESS

Published by the University of Pittsburgh Press, Pittsburgh, Pa., 15260
Copyright © 2011, University of Pittsburgh Press
Manufactured in the United States of America
Printed on acid-free paper
10 9 8 7 6 5 4 3 2 1

ISBN 10: 0-8229-6163-6
ISBN 13: 978-0-8229-6163-5

Cataloging-in-Publication Data is available from the Library of Congress

CONTENTS

ACKNOWLEDGMENTS

The idea for this edited volume originated in a research group on "Science in the Context of Application" at the Center for Interdisciplinary Research / Zentrum für interdisziplinäre Forschung (ZiF), University of Bielefeld, Germany. All but two of the contributors to this edited volume were fellows of the group. It is a pleasure to thank the ZiF and all members of this group for stimulating discussion and feedback. In addition, the editors would like to thank Johannes Lenhard for assisting us in various practical matters. Finally, Hans Radder would like to acknowledge the support for his contribution to this project provided by a fellowship of the Netherlands Institute for Advanced Study in the Humanities and Social Sciences in Wassenaar, the Netherlands.

SCIENCE TRANSFORMED?

Science after the End of Science?
An Introduction to the "Epochal Break Thesis"

ALFRED NORDMANN,
HANS RADDER, and
GREGOR SCHIEMANN

IN THE FEBRUARY 2008 ISSUE OF *Nature Nanotechnology*, physicist Philip Moriarty published a commentary that aims to reclaim academic science from postacademic science. Even though many of his readers are not at all familiar with the terms "academic" and "postacademic" science, Moriarty makes clear that the stakes are high. He is debating no less than the question whether it is still possible today to uphold an idea of science that values above all intellectual qualities like curiosity, creativity, and knowledge, and that does so for the sake of the public rather than the corporate good. At stake in reclaiming this idea of science is what might be called an "epochal break"—the idea that there has been a transformation in the relation of science, technology, and society so profound that our received notions of "science" have been superseded by something else.

It is telling that Moriarty's intervention appeared in a journal devoted to nanotechnology, which for some (for instance, Thomas Vogt, Davis Baird, and Chris Robinson [2007]) is the prime exemplar of a new age of technoscience. In the case of nanotechnology, Vogt, Baird, and Robinson argue, it is so utterly misleading to speak of "pure science" that it is actually morally bankrupt to pretend otherwise. Only those who openly acknowledge the technical, commercial, societal character of nanotechnological research can realize its potential to benefit humankind. Moriarty (2008, 61) responds: "It is the focus on market-driven wealth creation within publicly funded academic research, and not the distinction academics draw between 'pure' and 'applied' science, which is morally bankrupt." Another physicist, Richard Jones (2008, 448), comments on this exchange by remarking that scientists "for whom the traditional values of science as a source of disinterested and objective knowledge are precious" regard arguments for postacademic technoscience "as assaults by the barbarians at the gates of science."

Debates like this constitute one of the starting points of this edited volume. It is a debate about facts and about values. Has there been an epochal break or not? What happened to science as we knew it? And what does all this mean for science and society, for our intellectual traditions and the public good? Here, we first introduce the issue and the range of positions that have been adopted in the debate. In the second section we briefly summarize the chapters that make up this volume.

1. Introducing the Epochal Break Thesis

Almost every year serves as a banner year for science: 2009 was a case in point with the four-hundredth anniversary of Kepler's first two laws of planetary motion as well as Galileo's first use of a telescope for astronomical observations. Even more prominently, that year saw the celebration of the two-hundredth anniversary of Charles Darwin's birth and the hundred-and-fiftieth anniversary of his *Origin of Species*. Everyone recognizes the scientific accomplishments in all this—the advancement of knowledge toward a better understanding of the world, the conflict of science and religion, and a manner of inquiry that prizes critical thinking above all. Yet, even as we are celebrating these anniversaries and valorizing a certain image of science, we are expecting from contemporary research not primarily the discovery of truth but the solution of pressing problems—new ways to generate and store energy, cures for cancer and Alzheimer's disease, innovative ideas for sustainable economic development. Evolutionary biology, the neurosciences, and theoretical physics still command interest and curiosity, but the most prestigious research nowadays comes under the headings of nanotechnology, genetic engineering, biomedical research, or synthetic biology. So, when we celebrate Kepler, Galileo, and Darwin as great scientists, are they representatives of science as we value it today?

Answering this apparently simple question proves to be a difficult and controversial affair, and as Moriarty has demonstrated, there is a good deal at stake. Of course, one can quickly come up with symptomatic descriptions of changing conditions under which scientific research is undertaken—universities as patent holders, the computer as a powerful new tool, corporate sponsorship of research, and so on. However, this collection seeks to go beyond description. It also debates the meaning of these changes, since what is at stake is no less than a revered social institution that claims to provide an independent voice of reason for society to critically reflect upon itself. To abandon this institution would be tantamount to severing the alliance between science and the Enlightenment. Some argue that just this has happened in recent years. They maintain that there has been an epochal break that produced a profound reorientation of research practice. Others believe that there is nothing to worry about and that the situation today is not that different from the past. Yet others claim that this alliance has never really existed anyhow and that all-too-lofty views of science or of the Enlightenment never were anything but ideological.

This much is clear enough and not contested by anyone: science has never been free of interests and has always been conducted in a context of application. At least since the time of Francis Bacon (1561–1626), societies have looked to science to provide answers to their problems, to stimulate the economy, to inspire generally useful applications. Arguably, research scientists didn't live up to these expectations until sometime in the nineteenth century. These expectations were epitomized in the motto for the 1933 Century of Progress exhibition in Chicago: "Science finds, industry applies, man conforms." This slogan suggests a linear progression from scientific research to its technical applications and societal impacts. On this account, science enters a context of application only when it is very far along. Not until fairly recently has an awareness risen of rather more complicated interactions between science and technology. For example, we increasingly view the world around us as a product of science and technology and understand that science does not take its problems only from nature. Instead, many scientific problems arise from our reliance on the technological uses of science itself. These problems—like global warming, the toxicology of nanoparticles, or the exploitation of renewable energies—surface from complex interactions between social, technical, and "natural" factors. Science in the context of application is challenged to gain a new understanding and control of complexity—it cannot seek shelter in an idea of "pure science" or retreat to an imagined ivory tower.

This suggests that there are various ways for science and technology to interact in the context of application. Of the following three ways, the first two recapitulate familiar views of the relation, while the third may reflect a more troubling and more contemporary situation.

- Scientific research creates new technical capabilities that are then developed in engineering contexts. A prime example of this would be Heinrich Hertz's experimental and theoretical investigations of electrodynamics, which gave physical meaning to Maxwell's equations. A few years later, Guglielmo Marconi built on Hertz's findings to develop wireless telegraphy and thus prepare for the radio.

- Technological innovation gets ahead of scientific understanding and prompts research activity to attain comprehensive knowledge of its basic principles. The classic example is the construction of the steam engine, which proceeded mostly by trial and error without systematic understanding of the relation between work and heat. It was the steam engine itself and its successful operation that prompted scientists to study and finally understand this relation.

- Piecemeal research activities are commissioned to manage the complexity of sociotechnical systems with no particular expectation of comprehensive understanding. Take, for example, research to determine whether a certain salt mine is suitable for long-term storage of nuclear waste. This can involve a sizable interdisciplinary team of researchers for a number of years. What is here considered to be successful or conclusive research depends to a considerable extent on the informational demands of the decision-makers.

If recent developments force upon us a broad perspective on science and technology in the context of application, it is yet not broad enough. There is more that calls for our attention than the various relations of science and technology. The context of application is characterized not only by questions of use, by demands for theoretical understanding and public utility, and by the intended and unintended effects of scientific and technological innovation. The context of application is also a public sphere and a media culture, it is shaped by a variety of actors and institutions, by the pictorial representations of data and the resonance of visions.

The practical relevance of science, its great technological ambitions, its public appeal, and the heavy application pressure under which it operates today have prompted a flurry of analyses. A brief survey will show that they converge in the claim that science has undergone a profound methodological and institutional transformation during the past decades, perhaps an epochal break. With the flurry of analyses came a somewhat bewildering multitude of labels to designate the difference to traditional academic or theoretical science. Our quick overview will introduce some of the catchwords that are used to flag this transformation, though one should keep in mind that the following chapters are not so much concerned with the specific catchwords and labels but rather investigate the motives or causes that might have led so many to diagnose a profound transformation of science in the first place.

Most of the labels and diagnostic analyses in question originate in observations of the social and political conditions that influence science policy and research funding. In the 1990s, Henry Etzkowitz (2003; Etzkowitz and Leydesdorff 1998) began promoting the notion of a *triple helix* of *entrepreneurial science*, emphasizing how academia, industry, and government have become intertwined in the pursuit of research agendas. John Ziman (2000) complements this perspective with his discussion of *postacademic science* and the norms that guide it. According to Ziman, it is proprietary (rather than communal), local (rather than universal), authoritarian (rather than disinterested), commissioned (rather than original), and expert (rather than sceptical). Where Etzkowitz explores new opportunities for triple helical science policy, Ziman notes with a bit of alarm that the ethos of academic science is challenged by that of industrial or entrepreneurial science. Further studies have documented the profound effects of commercialization on university teaching and administration, and have related these phenomena to the rise of a neoliberal worldview and politics.

A mostly sociological and political characterization of the "new production of knowledge" has been offered also by Helga Nowotny, Peter Scott, and Michael Gibbons (2001), among others. They foreground a new social contract between science and society. Traditional or "mode-1 research" followed a trajectory of internally generated problems and procedures and was conducted in a setting where the pursuit of scientific questions was fairly well insulated and protected against immediate external interference and demands for accountability. This insulation is exemplified by large research laboratories and their closed walls behind which experiments are performed. This kind of research still exists today, of course, but it is being displaced by "mode-2 research," which is a more open undertaking that is characterized by a transdisciplinary orientation toward social, environmental, industrial, or medical problems. The fact that the boundary between science and society has become increasingly porous is seen as a cause for celebration because it suggests new opportunities for the social shaping of science and technology.

In a rather different setting arose the term "technoscience," which was introduced by Gilbert Hottois and popularized by Bruno Latour (1987, 1993) and Donna Haraway (1997). These rather more philosophical analysts of science do not use the term to claim that today's technoscience is radically dissimilar from previous science. What has changed is the way we look at science. For a long time, there was an effort to keep science as a quest for knowledge separate from technology as a way of changing our living conditions. This separation mirrors the effort to distinguish nature as a given, mind-independent reality from culture as the product of human action. But these attempts, Latour and Haraway agree, are futile, and under the label of technoscience we can now admit to that. The term fuses two words and it designates the ubiquity of hybrids. In technoscience, heterogeneous actors draw on conceptual and material resources to forge new kinds of entities, including technical artifacts. This perspective on science and technology, Paul Forman (2007) argues, coincides with postmodernity. In the modern age technology was viewed as applied science, while in postmodernity science is regarded as a kind of applied technology—its intellectual and physical control of phenomena depends on technology and a technological mode of thought. Again, these various thinkers evaluate technoscience rather differently. Latour and Haraway emphasize primarily that this new understanding enables new ways of acting and interacting, while Forman laments that science has become subservient to the realization of desired ends by any means necessary.

Forman does not introduce specific labels for the different ways of conceiving the relationship between science and technology. Though he attributes the current way of thinking to postmodernity, he does not speak of "postmodern science." That label has been used by others without catching on as of yet. For some, like Stephen Toulmin (1992), postmodern science is a program more than a reality. It is a kind of disunified science that recognizes a multiplicity of standpoints and respects local conditions. Others, like Jan C. Schmidt (2007), use postmodern science, or *nachmoderne Physik,* to designate research that draws on theories of complexity and self-organization rather than privilege isolable cause-effect relations. The identification, characterization, simulation, and "domestication" of particular highly complex phenomena resembles a "new natural history," as Arie Rip (2002) has pointed out.

Along some of these same lines, the term "postnormal science" has gained some prominence. It is associated primarily with Silvio Funtowicz and Jerome Ravetz (1993, 2001) and deals with scientific inquiry in high stakes situations where the disciplinary knowledge of normal science needs to be extended in various directions to cope with real-world complexities and the irreducible uncertainties that attend them. The production of new forms of ignorance in the course of scientific and technological development has been said by Ulrich Beck, Anthony Giddens, and Scott Lash (1994) to give rise to a "second modernity" or "reflexive modernization." This form of modernization mobilizes novel approaches to governance but also to the production of scientific knowledge in order to deal with the often unintended and unpredictable effects of moderniza-tion. In particular, it necessitates systematic reflection in order to cope with the risks of the uses of science itself.

These catchwords are by no means the only ways by which various authors seek to express what they perceive to be distinctive of much contemporary re-search. In the 1970s already, Gernot Böhme, Wolfgang Krohn, Wolfgang van den Daele (1973), and Wolf Schäfer (1983) spoke of "finalized science." Once the business of internal theory development has been finished, research needs to orient itself explicitly toward specific social or technical ends that are to be achieved. Much more recently, Peter Galison (2006) began speaking of an "en-gineering way of being in science" that is characterized by "ontological indiffer-ence," while Ann Johnson (2009) employs the notion of "research in a design mode." These terms capture the fact that many current research activities are more concerned with building or making than with knowing. Media theorists, art historians, and philosophers of modeling each from their own disciplinary perspectives ask whether there has been a major shift in the representational practices of science. And so, the list can be continued.

Some of these terms—"technoscience" and "mode-2 research" in particu-lar—will reappear throughout this book and to the astute reader, they may lack a proper definition. Indeed, more often than not, they are loosely descriptive of a phenomenon that remains to be fully understood. It is for this reason that these chapters seek out what it is that motivates all these various descriptions: What is the significance of the purported changes that draw so much attention? What, if anything, is new here? Ought we welcome this novelty or be troubled by it?

There is another reason why the chapters herein do not enter the thick of labels, adjudicating and comparing them one by one and one against the other. Rather than become entangled by them, it is important to reclaim the critical distance that allows us to ask what is at stake in these various descriptions and redescriptions of research practice. Only when this distance is maintained, the various accounts of the distinctiveness or novelty of contemporary research do not end up as self-fulfilling prophecies. Such a pattern can indeed be observed when, for example, the diagnosis of "triple-helical" or "mode-2 research" is taken up by science observers, foresight analysts, or policy makers, and when it becomes institutionalized in transdisciplinary funding practices that set out to enable a more effective technology transfer. Soon enough, philosophers, historians, and sociologists of science have themselves become caught up in the context of application—engaged in *sozialwissenschaftliche Begleitforschung*,

"ELSA-studies," or the facilitation of responsible development of emerging technologies. Against this background, this edited volume attempts to reclaim a critical perspective on contemporary developments.

Finally, the problem of definition also applies to the term "epochal break." Whether one ends up proclaiming such a break depends on what one imagines an epochal break to be. Are we taking as our model the epochal break between the medieval "dark" age and the renaissance with its light of reason? On this model, the burden of proof would be quite high, but a case might be made, for example, by arguing that we are moving from a period of disenchantment, rationalization, and intellectualization to an age where the technoscientific world of our own creation becomes an enchanted, magical place. Less ambitious claims confront difficulties of their own. We might take as our model, for example, a so-called Kuhnian scientific revolution or paradigm shift, though paradigm shifts typically occur within physics, chemistry, or other disciplines, whereas "technoscience," "mode-2 research," and the like refer to changes that affect all disciplinary research across the board. However, a case might be made for the Kuhnian model by speaking about the recent emergence of new disciplines with new paradigms and problems somewhat along the lines of the emergence of molecular biology many decades ago. The so-called Hacking revolutions provide a third model. These refer to a conceptual or technical innovation that can mark a point of no return. Similar to the "probability revolution" of the eighteenth and nineteenth centuries, for instance, the introduction of desktop computing and simulation modeling may have changed forever and for everyone the rules of the game of explanation and understanding, predicting and controlling the world—irrespective of whether an individual researcher employs such models or not.

These three models do not exhaust the many ways of speaking about an epochal break. For example, we saw Forman (2007) modeling the epochal break on the transition from modernity to postmodernity, by which he does not mean a break on the level of practice but on the level of ideology, interpretation, or cultural prestige. Media theorists refer to the epochal transition from analog to digital imaging, which severs the traditional causal chain from the original to its representation and allows any kind of data to be rendered in any number of visual forms. Yet another model is associated primarily with Michel Foucault's notion of "epistéme" and a shift in the order of discourse—that is, in the presuppositions that accord power and efficacy to certain kinds of knowledge. In light of these various meanings of core concepts, the question this book asks does not have a straightforward answer. Is the manner in which knowledge is produced business as usual, or do changing relations of science and technology signify an epochal break? The different ways of approaching this question illustrate the many ways of reflecting upon our age and the contemporary significance of science in and for society.

2. Debating the Epochal Break Thesis

The chapters included in this edited volume address the epochal break thesis from a variety of disciplinary backgrounds, including philosophy and history of science, social studies of science and technology, and cultural and media studies of science and technology. The subsequent chapters are divided into two groups. The first group of chapters seeks to adjudicate the proposed claim of an epochal break as a whole. It opens with several chapters that start the discussion by providing strong views in favor of or against the idea that during the past decades there has been a profound reorientation of the scientific enterprise. The authors in the second group approach the thesis from more specific perspectives. They foreground certain concepts, single out specific technical developments, or consider particular practices and contexts of application. These specific concepts, technologies, and fields of practice serve as a testing ground for the larger thesis.

Alfred Nordmann interprets the epochal break as a shift from the scientific enterprise to the regime of technoscience. Characteristic of the scientific enterprise is that representing and intervening—nature and culture, science and technology—are taken to be distinguishable. The critical point of distinction, then, is that for technoscience this purification (of nature from culture and so on) is no longer possible and no longer required. Methodologically, Nordmann argues in his chapter that this thesis cannot simply be inductively proven in an empiricist way but requires the reasoned adoption of a specific vantage point. However, the thesis can and should be empirically articulated and supported by specifying in detail how both the separation and the conflation of science and technology work out in actual scientific and technoscientific practices.

In his chapter Gregor Schiemann addresses the issues through the notion of a scientific revolution and claims that at present we are not witnessing a new scientific revolution. Instead, Schiemann argues that after the so-called Scientific Revolution in the sixteenth and seventeenth centuries, a caesura occurred in the course of the nineteenth century that constituted a departure from the early modern origins of science. This change was characterized by the loss of certainty on the part of the scientists, by the steadily increasing importance of scientific communities (rather than individuals), and by the systematic intertwinement of scientific and societal development. As to present science, Schiemann admits that important changes have occurred, but he denies the conflation of nature and culture: even the OncoMouse is a natural organism, though a seriously damaged one.

Martin Carrier then offers a radical counterpoint to the epochal break thesis. Instead of claiming that science is *no longer* interested in a theoretical understanding of the world and that it is *now* in the service of ambitions to make and remake the world, one should realize that modern science has always pursued the latter ambition but is only now able to deliver on its promises. The notion of science as a theoretical enterprise that is ultimately interested in truth was offered as the royal road to the larger aim of utility and benefit. Throughout its history, and still today, theoretical understanding was required for the advancement of technical goals. In this sense modern science was and continues to be an epistemic enterprise. As it delivers on its promise, however, there is a change at the ontological level: many or most of the objects studied in present-day sciences, such as nanotubes or nonsteroid anti-inflammatory drugs, are not part of an untamed nature but the result of artificial human creation.

From his historical perspective, Cyrus Mody is skeptical of any grand claim about epochal breaks. Thus he argues that key aspects of current technoscience can be found in postwar nuclear physics (Alvarez and nuclear weapons), in early-twentieth-century physical chemistry (Langmuir and lightbulbs), in the study of electricity and magnetism in the second half of the nineteenth century (Kelvin and the telegraph), and even in seventeenth-century mechanics (Galileo and the Medicis). Mody points out that announcements of epochal breaks are often interested and do have real consequences—for instance, in terms of funding policies. He concludes that the scholarly study of science should be wary of epochal break talk and keep firmly in mind that one particular vantage point is as good as any other.

A different but equally critical view is taken by Mieke Boon and Tarja Knuuttila. They discuss the alleged divides between representing and intervening and between basic and applied research. Here representation and basic research would be typical of mode-1 research, while intervention and applied science would characterize mode-2. Because models are multifunctional epistemic tools, the uses of which may include both representing and intervening, and since modeling is, and has been for long, the core of the natural and in particular the engineering sciences, the epochal break between mode-1 and mode-2 research collapses. Like Carrier, Boon and Knuuttila suggest that there has never been mode-1 research. What has changed, though, is the political rhetoric that exploits mode-2 talk for legitimizing short-term accountability and commodification.

Hans Radder's chapter exemplifies a more differentiated approach. It points out that human intervention in nature has been an essential dimension of the experimental tradition since its inception in the Renaissance. It also questions the philosophical and empirical claims about the absence of purification work in present-day science. However, Radder argues, this does not mean that recent science is business as usual. In particular, he points to the rise of important, nonlocal patterns, such as the significance of the external validity of scientific knowledge and methods and the commodification of academic research. The identification and explanation of such nonlocal patterns requires a subtle and reflexive methodology, which precludes the positing of Great Divides and acknowledges the unavoidable value-ladenness of this type of historical-philosophical research.

The chapter by Andrew Jamison provides a brief account of the changing contexts of science and technology since the 1940s. He distinguishes a mode-1 phase of disciplinary little science (before World War II), a mode-1½ phase of multidisciplinary big science (the 1940s through the 1960s), and a mode-2 phase of transdisciplinary technoscience (since the 1970s). In response to the criticisms of commercialized technoscience, Jamison claims that we cannot return to a mode-1 science that is no longer meaningful, and he tentatively argues for a mode-3 phase as a desirable synthesis of traditional and commercial research cultures. As for the idea of an epochal break, Jamison speaks more carefully of "changing contexts," while the notion of a mode-1½ phase suggests a transitional process.

In the following chapter Chunglin Kwa admits that nowadays science is primarily seen as technoscience. But although the idea of technoscience seems to imply a primacy for technology with respect to science, he offers an alternative model for the interaction between science and technology. Building on A. C. Crombie's work on the six styles of scientific thinking, Kwa proposes design, in the literal sense of *disegno* or drawing, as the core of a technological style. In technoscientific practice this technological style may forge a variety of alliances with other scientific styles. The approach is illustrated by analyzing several alliances of airplane design with instances of other scientific styles. The idea of alliances entails that technology, or a technoscientific enterprise, is combined with, and hence has not replaced, science or a scientific enterprise.

The chapters in the second part of the book address more specific aspects of the epochal break thesis. This does not mean that they deal with minor details, because they do discuss substantial patterns or trends in the history of recent science. Astrid Schwarz and Wolfgang Krohn begin by showing that the concept of "experiment" has undergone a major shift that prepared or accompanied a more general reorientation of the relationship between science and society. As long as experiments were confined to the laboratories of the classical natural sciences, this relationship was defined by a clear separation of spheres. The increased significance of field experiments, which do not rely on the isolated space of the laboratory but transform a field site or social setting into a laboratory of sorts, prepared an understanding of large-scale "real-world experiments" that serve as sites for social learning.

Valerie Hanson's chapter focuses on the characteristics of the strongly increased uses of digital imaging in science. An important new aspect of these visualizations is that they enable the viewers to interact with the objects of study. Making such images—for instance, of molecules and molecular processes in chemistry—becomes an exploratory, experimental procedure. Furthermore, this technology entails new rhetorical strategies for disseminating knowledge, both among the scientists themselves and among the general public. Although digital imaging possesses several features that are different from analog visualizations, Hanson emphasizes that it is not fully novel. Hence her conclusion is that the effects of digital media on scientific practice are a matter of intensification, not of full transformation.

In contrast, Angela Krewani considers a fundamental change in the representational practice of the sciences. With digital media technologies, she explains, any set of data can be rendered as an image and any image can be decomposed into a data-set. She shows how analog and digital media technologies condition behavioral attitudes that simultaneously constitute the scientific object and the scientific observer of that object. Even though digital imagery is sometimes used merely to emulate analog imaging techniques, this should not detract us from the fact that digital technologies presuppose and enable a different kind of interactivity and thereby also a different spatial relation between objects and subjects of research.

The recent developments in robotics and human-robot interaction are the topic of Jutta Weber's contribution. She documents the shift, in the late 1980s and the 1990s, from robots as industrial tools to a focus on personal-service robots. The latter are claimed to be not just devices but rather artificial creatures, possessing cognitive, affective, and communicative capabilities. Up to now, the cultural significance of humanoid robotics is primarily symbolic: it represents glamour science, it provides edutainment, and it shows that science can be fun. At a more basic level, however, human-robot interaction challenges fundamental assumptions about what it is to be human. Regarding the theme of this book, humanoid robotics should be seen as a full-blown technoscience. It is neither about representing nor about intervening, but rather about constructing and reconstructing hybrid cyborg worlds.

In his chapter, James Robert Brown observes that one aspect of the science-technology relationship has been underexposed thus far—namely, the erosion of scientific quality under the influence of technological and commercial interests. With the help of detailed examples, he demonstrates that such erosion has occurred, and still occurs, in medical science. Furthermore, the positive impact of pharmaceutical research is much smaller than often claimed. Solving these problems, he argues, requires some radical changes: not merely stricter regulation and public control of clinical trials, but also the abolishment of all intellectual property rights and a policy of full public funding. Brown concludes that although there has been an (undesirable) epochal break in medical research, this cannot be generalized to other areas, such as high-energy physics.

Finally, Ann Johnson and Johannes Lenhard discuss the rise of cheap and widely available personal computers during the 1990s and argue that this development has led to "a new culture of prediction." The performance of computational models is often opaque, and usually the models themselves cannot be interpreted as a realistic representation of the way the world is. What such models do possess, however, is a substantial predictive power. Philosophers of science often focus on explanation, but the more prominent role of prediction requires an independent historical and philosophical analysis and assessment. Johnson and Lenhard conclude that prediction is the key to technological control. Moreover, the increasing impact of opaque computational modeling demonstrates that such control can be achieved without in-depth theoretical understanding.

The epilogue, written by Hans Radder, is of a different nature. It does not argue for or against the epochal break thesis or some of its aspects. Instead, it attempts to extract from the preceding chapters a number of issues that can be expected to remain the focus of sustained research and debate. These "sticking points" include historiographical questions of how to support a comprehensive historical thesis like the epochal break thesis, ontological and epistemological issues about the nature and development of the sciences, empirical and theoretical accounts of the role of old and new methodologies, social-scientific inquiry into the relationships between science, technology, and the wider society, and normative concerns about the sociocultural roles of science and technology.

This introductory overview offers no more than a sketch of the main points of the chapters. In fact, the chapters cover a richer set of relevant subjects and a more subtle variety of arguments and positions than can be covered in this introduction. To discover this richness and subtlety, please read on and see for yourself.

REFERENCES

Beck, Ulrich, Anthony Giddens, and Scott Lash. 1994. *Reflexive Modernization: Politics, Tradition, and Aesthetics in the Modern Social Order*. Stanford: Stanford University Press.

Bensaude-Vincent, Bernadette. 2009. *Les vertiges de la techoscience: Façonner le monde atome par atome*. Paris: La Découverte.

Böhme, Gernot, Wolfgang Krohn, and Wolfgang van den Daele. 1973. "Die Finalisierung der Wissenschaft." *Zeitschrift für Soziologie* 2: 128–44.

Carrier, Martin, and Alfred Nordmann, eds. 2010. *Science in the Context of Application*. Dordrecht: Springer.

Elzinga, Aant. 2004. "The New Production of Reductionism in Models Relating to Research Policy." In *The Science-Industry Nexus: History, Policy, Implications*, edited by Karl Grandin, Nina Wormbs, and Sven Widmalm, 277–304. Sagamore Beach, Mass.: Science History Publications.

Etzkowitz, Henry. 2003. "Innovation in Innovation: The Triple Helix of University-Industry-Government Relations." *Social Science Information* 42: 293–337.

Etzkowitz, Henry, and Loet Leydesdorff. 1998. "The Endless Transition: A Triple Helix of University-Industry-Government Relations." *Minerva* 36: 271–88.

Forman, Paul. 2007. "The Primacy of Science in Modernity, of Technology in Postmodernity, and of Ideology in the History of Technology." *History and Technology* 23: 1–152.

Funtowicz, Silvio O., and Jerome R. Ravetz. 1993. "The Emergence of Post-Normal Science." In *Science, Politics, and Morality: Scientific Uncertainty and Decision Making*, edited by René von Schomberg, 85–123. Dordrecht: Kluwer.

———. 2001. "Post-Normal Science: Science and Governance under Conditions of Complexity." In *Interdisciplinarity in Technology Assessment: Implementation and Its Chances and Limits*, edited by Michael Decker, 15–24. Berlin: Springer.

Galison, Peter. 2006. "The Pyramid and the Ring." Presentation at the conference of the Gesellschaft für analytische Philosophie (GAP), Berlin.

Gibbons, Michael, Camille Limoges, Helga Nowotny, Simon Schwartzman, Peter Scott, and Martin Trow. 1994. *The New Production of Knowledge: The Dynamics of Science and Research in Contemporary Societies*. London: Sage.

Haraway, Donna. 1997. *Modest_Witness@Second_Millennium*. New York: Routledge.

Ihde, Don, and Evan Selinger, eds. 2003. *Chasing Technoscience*. Bloomington: Indiana University Press.

Johnson, Ann. 2009. *Hitting the Brakes: Engineering Design and the Production of Knowledge*. Durham, N.C.: Duke University Press.

Jones, Richard. 2008. "The Production of Knowledge." *Nature Nanotechnology* 3: 448–49.

Latour, Bruno. 1987. *Science in Action: How to Follow Scientists and Engineers through Society*. Cambridge: Harvard University Press.

———. 1993. *We Have Never Been Modern*. Cambridge: Harvard University Press.

Moriarty, Philip. 2008. "Reclaiming Academia from Post-Academia." *Nature Nanotechnology* 3: 60–62.

Nowotny, Helga, Peter Scott, and Michael Gibbons. 2001. *Rethinking Science: Knowledge and the Public in an Age of Uncertainty.* Cambridge, Mass.: Polity.

Radder, Hans, ed. 2010. *The Commodification of Academic Research: Science and the Modern University.* Pittsburgh: University of Pittsburgh Press.

Rip, Arie. 2002. "Science for the Twenty-First Century." In *The Future of the Sciences and Humanities: Four Analytical Essays and a Critical Debate on the Future of Scholastic Endeavour,* edited by Peter Tindemans, Alexander Verrijn-Stuart, and Rob Visser, 99–148. Amsterdam: Amsterdam University Press.

Schäfer, Wolf, ed. 1983. *Finalization in Science: The Social Orientation of Scientific Progress.* Dordrecht: Reidel.

Schmidt, Jan C. 2007. *Instabilität in Natur und Wissenschaft: Eine Wissenschaftsphilosophie der nachmodernen Physik.* Berlin: de Gruyter.

Slaughter, Sheila, and Gary Rhoades, eds. 2004. *Academic Capitalism and the New Economy: Markets, State, and Higher Education.* Baltimore, Md.: Johns Hopkins University Press.

Toulmin, Stephen. 1992. *Cosmopolis: The Hidden Agenda of Modernity.* Chicago: University of Chicago Press.

Vogt, Thomas, Davis Baird, and Chris Robinson. 2007. "Opportunities in the 'Post-Academic' World." *Nature Nanotechnology* 2: 329–32.

Weingart, Peter. 1997. "From 'Finalization' to 'Mode 2': Old Wine in New Bottles?" *Social Science Information* 36: 591–613.

Ziman, John. 2000. *Real Science: What It Is, and What It Means.* Cambridge: Cambridge University Press.

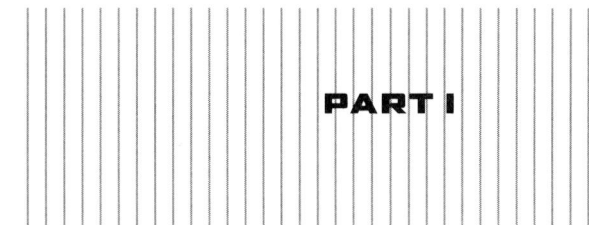

PART I

The Age of Technoscience

ALFRED NORDMANN

MODE-2 RESEARCH, POSTACADEMIC SCIENCE, technoscience, postnormal science, new natural history, entrepreneurial science—all these various labels speak of more or less profound changes in the organization of research. Do these changes amount to an epochal break that transforms scientific knowledge production as a whole? The theories behind each of these designations do not offer straightforward answers to this question. If a new kind of commissioned research enters the scene in the late twentieth century, this might leave most of the sciences unaffected. And if today's research practices defy notions of "pure research" or "basic science," and if they thereby open our eyes to the rich interactions between science, technology, and society, this might lead us to see these rich interactions also in the past. All that has changed, some would argue, is how we appreciate scientific practice, but the business of science is as complex as it has always been.

Instead of reviewing various accounts of past and current research practice, I want to make a case for an epochal break between the scientific enterprise and the regime of technoscience. However, to make a case is different from settling a matter of fact. Rather than decide whether or not scientific knowledge production has changed as a whole, I want to show in what sense it is adequate, illuminating, even important to consider the various diagnosed changes in terms of an epochal break. Doing so is not a neutral exercise but motivated by concern for the scientific enterprise. From the point of view of science and how it understands itself, hardly anything could be as dramatic as the shift to a technoscientific mode of research. From the point of technoscience, in contrast, the whole history of science and engineering research has always been technoscientific. In a final, apparently dialectical twist to my argument, I therefore argue that those who deny the epochal break have happily settled into the age of technoscience, while those who see an era coming to an end are those who care about science and its deep connection to modernity and the Enlightenment project.

I.

What is an epochal break? Surely, it is not a moment in time when, suddenly, everything changes and the world is becoming a different place. Some have argued that World War I or the nuclear destruction of Hiroshima were such moments. Others cast doubt on such ruptures but see them as salient moments that grew out of the past and beyond which much continued as it was. This is true also for the epochal break that matters most and that shaped our very idea of an epochal break—namely, the transition from a medieval or premodern world to the modern world. As Hans Blumenberg (1976), in particular, has pointed out, there would be no modern world without the assumption of an epochal break—to be modern is to distinguish oneself from those who came before, to respond to the seriousness of one's age, to relate oneself to the demands of the day. And yet the history of modernity is full of uncertainty and controversy about the precise time and place, the extent and significance of the transition from medieval to modern times. Still, to be modern is to frame one's own place in the world historically, part of a movement from one era to the next, each with its own character and destiny. Even as the moderns remained profoundly unsure how they could and should distinguish themselves, they liberally proclaimed epochal breaks, most prominently in the philosophy of Hegel or in the case of Goethe, who declared the beginning of a new era after witnessing an all-but-forgotten battle in one of the countless wars between the Germans and the French.[1] The notion of "epoch" or "era" became an instrument of the moderns to reflect upon themselves, their place in history, the distinctiveness of their times, and thus of their calling.

In recent years, one of the preeminent philosophers of technoscience, Bruno Latour (1993), has argued that we have never been modern. His claim does not contradict Blumenberg's but complements it: modernity presupposes that one can distinguish the modern self from that of the dark ages, that one can distinguish culture from nature, science from technology, this era from another. Since we have never quite succeeded in establishing and fortifying these distinctions, we have never been modern. And yet it is characteristically modern to engage in such work of purification—that is, to engage in the work of distinguishing oneself, of attributing blame either to nature or to human intervention. This is Blumenberg's point: there is no compulsion from facts or principle that would force anyone to see an epochal break here or there; but to see an epochal break is tantamount to accepting one's historical destiny or mission, and this is what moderns do.

This is also how one should understand my attempt to identify the epochal shift from the scientific enterprise to the regime of technoscience. I begin by acknowledging that I cannot compel my readers to see this shift. They can see it only if they willingly follow me to the particular vantage point from which it becomes visible. And by following me, they will see two projects: a historical project that is called "the scientific enterprise" and the project of technoscientific innovation, which turns out to lack a historical mission and which thereby lies beyond modernity and its obsession with epochal breaks.

II.

To see the epochal break in question requires the adoption of a vantage point from which it becomes visible. What is this vantage point? It consists in the proper distance to scientific practice, one that is neither too remote from nor too close to what is happening on the ground. Looking very closely at the particulars of research practice teaches us first and foremost that there is neither "science" nor "technoscience" but a multitude of ever-shifting disciplinary formations that are guided by specific epistemic values, experimental, observational, and representational practices, patterns of explanation and intervention. From this perspective, nothing could be more misleading than to speak of a transition from "science" to "technoscience." It is wrong even to posit a monolithic and idealized notion of "science" in the first place. Instead, there was and is a multiplicity of sciences. Some of them are strongly oriented toward the demands of practice and closely aligned with what is now termed technoscience. Others have fashioned themselves after an unattainable ideal of "pure" or "basic" science as an unfettered search for truth. Indeed, no epochal break thesis should deny this multiplicity of the sciences and happily none does.[2]

It is less clear what one can learn from a rather remote perspective on this multiplicity of sciences. It produces useful generalities. It is useful in that it provides a conceptual frame that is broad enough to accommodate very heterogeneous practices. It is a frame within which one can differentiate disciplinary particularities. For example, one might say that all sciences seek an understanding of mechanisms by producing models that can be adapted to local phenomena. One might add to this that they do so even if their main purpose is not to seek understanding but to make things work. Because this general level of description is meant to subsume the specific differences between various modeling techniques, types of mechanisms, or notions of understanding, it is easy to see that it will also subsume the difference between sciences and technosciences. Significantly—and, as such, radically dissimilar from earlier attempts to provide a unified characterization of the sciences—the general description in terms of models and mechanisms avoids any attribution of a historical mission to research practice.[3] The business of science is to engineer a fit between the phenomena and some set of conceptual tools. Whether we look at theoretical and the so-called applied sciences, they are all engaged in this "business as usual"—and this unifying description already expresses a technoscientific point of view according to which all research is always in the business of designing a proper fit between general laws or theoretical ideas, models, and the phenomena and processes in the real world.

Since an epochal break is invisible from both these vantage points, it requires a third one to make it visible. It is a perspective from which one does not see just the multiplicity of disciplines and practices, and from which one does not make out merely the technical features of the general business of explanation and understanding. Instead, it is the vantage point from which one can see "the scientific enterprise" with its historical mission and then "the regime of technoscience."

III.

To see "the scientific enterprise" is something other than seeing "the sciences" or "modeling," "specifying mechanisms," "(scientific) explanation," or the like. The term "scientific enterprise" is on a par with terms like "the Enlightenment project" or "modernity" and indeed closely related to them. As with modernity and the Enlightenment, we might have a hard time knowing just when and where the scientific enterprise began and whether it ended. But we are nevertheless quite confident that it did not exist everywhere at all times. It is the name for a common pursuit that orients the various sciences and influences their self-definition. It suggests that, separately or together, all the different ways of knowledge production contribute to a historical process that, citing Max Weber (1919), might be referred to as a process of rationalizing or intellectualizing the world.

Taking a further cue from Weber, the scientific enterprise along with the

Enlightenment project are committed to an unending quest for truth and thus obliged to a future state of knowledge. We do not live in an enlightened age, Kant wrote, but in an age of Enlightenment, and this sentiment was echoed by countless philosophers until nearly the end of the twentieth century. In Karl Popper's philosophy of science, for example, it reappears as the edict that in the name of truth we formulate and criticize mere hypotheses but that the truth itself is an unattainable good that we can approximate, at best. What I call the historical mission of the scientific enterprise is its orientation to truth, where "truth" might be taken as its *telos*, as its regulative ideal, as a correlate to a certain kind of intellectual control, or as the original for which weaker notions of empirical adequacy or reliability serve as proxy. Whether it proceeds in a linear and cumulative manner or by way of detours and upheavals, the scientific enterprise advances an ever more detailed and pervasive determination and rational control of reality. If the complete determination of reality is tantamount to the truth—an idea that one finds in such diverse philosophers as Charles Sanders Peirce and the early Ludwig Wittgenstein—the scientific enterprise progressively approximates that truth. And to this determination of reality, to intellectualization and rationalization of the world contribute also those endeavors that are not dedicated to truth-for-its-own-sake or to a theoretical description of the world. Louis Pasteur, Justus von Liebig, Adolphe Quetelet, and Claude Bernard cannot be dissociated from the scientific enterprise and the Enlightenment project just because they developed tools and practices in the service of humankind.

The particular sciences are oriented toward the scientific enterprise each in their own way. Some assume the vanguard with a grand and arrogant gesture, defining a hierarchy of the sciences that sees them at the top. The more exclusively and directly it is committed to the scientific enterprise, the more basic or fundamental is the research in question. It is "pure" science precisely because its historical mission is unadulterated truth-seeking. All this may hold especially for theoretical physics and evolutionary biology. Other sciences put themselves humbly in the service of humankind, yet others plod along, confident and content to toss out a new theoretical challenge, an interesting discovery here and there—and the occasional inferiority complex of the "applied sciences" is just one way of acknowledging and reinforcing the need to coordinate one's self-definition with the overall ambitions of the scientific enterprise. As long as established fields of inquiry and practice—from sociology and engineering to nursing—seek to become "sciences," they are claiming that they, too, can systematically contribute to the overarching scientific enterprise and historical mission of rationalization, intellectualization, or Enlightenment.

To be sure, the orientation of the sciences to an overarching scientific enterprise is not without consequence for the epistemic values that govern various research practices. This is most obvious perhaps for conceptions of objectivity. It may be fair to generalize that scientific (as opposed to technoscientific) notions of objectivity are framed in terms of history. Knowledge is said to be objective if it endures and if it can be separated from the historical contingencies of its discovery and formulation. The production of objectivity with its concern to separate lasting, if not eternal, truths from changeable contexts and opinions is yet another instance of the previously cited work of purification. This work of purification also requires that one can distinguish between immutable aspects of nature and the human interventions that are required to make these aspects visible in contexts of experimentation and observation. One way of doing so is to regard nature as a collection of dispositions or latent behaviors that are prompted by researchers to become manifest and observable. A laboratory experiment, for example, serves as a trigger that stimulates the exhibition of a certain reaction or behavior. Though humanly or technically induced, the manifestation is nothing but nature's own doing—the experiment prompts nature to reveal itself. This conceptualization of the experiment is yet another example of the work of purification—it separates out the evidence for timeless scientific truth from the contingent particulars of human and technological intervention.[4]

The notion of the "scientific enterprise" is meant to serve as a middle term between the many particular sciences with their varied concepts of science and objectivity and the most general notions of how humans forge an agreement between their thoughts and the real world. Each scientific discipline may have its own paradigm, and within each discipline there might be scientific revolutions that involve paradigm shifts. In and of themselves, however, these do not constitute epochal breaks. Accordingly, Blumenberg (1976, 16) characterizes paradigm shifts as "a surrender of basic assumptions and the introduction of new elementary suppositions, which get rid of a desperate situation but do not necessarily rupture the identity of the movement of knowledge that had culminated in that situation." The scientific enterprise is that overarching movement of knowledge.

IV.

So much for the scientific enterprise. It is time to turn to the regime of technoscience, first to characterize it and then to reflect on the epochal break and the discontinuities that are signaled by it. As for its characteristics, many of them emerge from the contrast with the scientific enterprise. The regime of technoscience is not a new paradigm within the traditional movement of knowledge but reorients the various old and new ways of knowledge production. This different orientation has been associated with postmodernity especially by Paul Forman (2007, 2010). But rather than valorize postmodernity as a new epoch with its own historical mission and overarching movement of knowledge, it is sufficient to characterize it simply as the abandonment of the work of purification and of received notions of objectivity. Indeed, in the age of technoscience, the work of purification is no longer possible and no longer required. This can be shown for an "in silico" experiment that is performed in a computer model or for research on a genetically engineered laboratory model. In both cases it becomes impossible, as a matter of fact, to conceptually determine where human intervention ends and the purely natural process begins. Because these experiments serve mostly to demonstrate practically achieved control of the phenomena, there is also no need to determine this—the achievement stands on its own and validates itself.

To be sure, both aspects of this definition demand extensive substantiation. They imply two claims about changes in the practice and orientation of knowledge production: that the separation between nature and technology, between representation and intervention, is no longer possible, and that this separation is no longer required. With these two claims the epochal break thesis becomes an empirical thesis with its definite standards of evidence. Though many of the arguments for mode-2 research, the triple helix, postnormal science, and so on serve to corroborate these claims, the thesis cannot be fully supported here. More pressing for the purpose at hand is to spell out what is meant by saying that the work of purification has ceased and that objectivity is now conceived in an ahistorical manner.

The very term "technoscience" testifies to the abandonment of the attempt to separate "science" and "technology." Though the two have commingled in research practice throughout the history of science and the history of technology, the previous section introduced one of the conceptual devices by which it was relatively easy to separate them out in the mind and assign each to its own domain. Technology was said to provide active interventions that serve as triggers or stimuli in an experiment, whereas the response to these stimuli was thought to be an aspect of nature and as such subject to passive observation and scientific inquiry. The separation between technology and science thus mirrors the separation—in the mind and only as a result of the conceptual work of purification—between practical intervention in the world (technology) and theoretical representation of the world (science). It thereby also mirrors the distinction between on the one hand contingent events that are produced by experimenters, and on the other hand the immutable properties and processes of nature that reveal themselves in response to arbitrary technical interventions. It is entirely contingent whether two substances are brought in contact with one another during an experiment, but their chemical reaction is a matter of natural lawfulness. None of these distinctions applies to technoscientific research. The objects of research cannot be separated from the technical interventions required to produce, maintain, or observe them. A carbon nanotube or a genetically engineered laboratory animal exhibit properties and processes that are themselves engineered—their relevant dispositions are aspects simultaneously of nature and culture.

When there are no distinct immutable features of nature that are opposed to technology and culture, there is also no historical movement of thought that sets out to converge on these features and that seeks objectivity by gradually controlling for all sources of subjectivity in the course of producing intersubjective agreement about the best possible representation. Instead, technoscientific objectivity results quite literally from objectification—that is, from the transformation of available knowledge, technology, and skill into material processes and things. Instead of seeking time-independence, technoscientific objectivity requires independence from place. Laboratory constructions need to become delocalized to achieve the robustness required for their survival in the outside world. Technoscientific objectivity thus serves the so-called knowledge society as it draws on theoretical and instrumental tools to solve identified problems. This is a matter of innovation rather than progress. Accordingly, the technoscientific regime is not obliged toward a future state of true and proper knowledge. For technoscientific innovation, the future is merely a repository of technical possibilities that await to be realized. In the meantime its task is framed entirely by the present as one of matching up supply and demand—technoscientific capabilities and the societal or environmental problems that call for technical solutions.

The term "delocalization" was discussed, perhaps introduced, by Peter Galison (1997). Galison also speaks of "ontological indifference" to characterize technoscientific research (Galison 2006; see also Daston and Galison 2007, 393, 414). Engaged in the work of purification, the scientific enterprise must be deeply concerned with what is and what is not an aspect of nature and what is therefore subject to scientific representation and truth-seeking. Ontological indifference is indifferent to purification. Technoscientific research proceeds in a design or engineering mode, and the hallmark of good technoscience is the acquisition and demonstration of basic capabilities and, beyond that, the creation of robust technical systems. Clearly, the achievement of these capabilities and the robustness of these systems do not require for their effectiveness that the contributions of nature and human craft can be disentangled.

V.

This chapter posits an epochal break between the scientific enterprise with its historical mission and the regime of technoscience and its conquest of space. In particular, this amounts to the claim that the various scientific endeavors are oriented no longer toward the progressive ideals of the Enlightenment project, but instead to the innovative production of a local fit between available modeling tools and the phenomena, between achieved capabilities and recognized problems. The empirical criterion for this shift is the abandonment of the futile and nevertheless ceaseless work of purification that characterizes modernity.

The significance of this break can hardly be overstated. And yet this break does not imply radical discontinuities in research practice. One continuity arises from the observation (see section II earlier in this chapter) that science has always been concerned to produce a fit between available modeling tools and the phenomena. Indeed, such a description of practice might yield a lowest common denominator for scientific and technoscientific research. By the same token, it involves an impoverished description especially of the scientific enterprise (the proponents of technoscience are rather more willing to embrace it). Scientists who are working in the lab to produce such a fit but who are also concerned to prove themselves as "scientists" are producing this fit in pursuit of bigger objectives and in order to contribute to a larger project of Enlightenment.

Another dimension of continuity is that the regime of technoscience would not be possible if it could not draw on the knowledge and skill acquired in pursuit of the scientific enterprise. Indeed, the continuous progress of representational capabilities may well have given license to their use beyond representational purposes for a kind of substitutive, qualitative, immersive, and interactive mode of reasoning about technoscientific complexities (Nordmann 2006). Yet another aspect of continuity takes us beyond the philosophical interests of these remarks into the realm of sociology. The technoscientific regime may have displaced the scientific enterprise in the rhetoric of funding, of justification, and of framing research, but it has not displaced it in the education of physicists or biologists. Many young researchers are introduced to the "life of science" and then find themselves operating in a world of "strategic research" that is shaped by very different values. Indeed, the presence and reality of the technoscientific regime calls attention to itself by way of ambivalence—that is, by way of research that comes from one place and finds itself in quite another, and that learns to play a new game with many of the old rules.

On the side of the researchers or of the subjects, then, there is no clear division between the age of science and the age of technoscience. The epochal break runs right through them and is frequently experienced as an inner conflict. On the side of the objects of research, however, the division appears much clearer.[5] For example, the objects of technoscience do not have fixed and definite substantial natures but are mere potentials. Substantial natures determine what something is—a stone is hard as a rock. Considered as a mere potential, the stone is what it might become—that is, a momentary configuration of atoms and molecules that could be turned into just about anything else. Similarly, the computer's ability to solve humanly intractable equations affords not just an improvement of extant modeling techniques but is radically discontinuous with the past. Where models used to mediate between theory and reality and thus highlighted the distance between theory and reality, theory now offers algorithmic building blocks to construct a substitute reality that is hardly less complex than its original. And so, this list of discontinuities may be continued to include the role of conservation laws, the very idea of propositional knowledge that takes the shape of theories and hypotheses, or the disrepute and prominence of similarity and likeness in evidentiary reasoning.

VI.

In light of the various continuities and discontinuities in the shift from the scientific enterprise to the regime of technoscience, the question arises how it came to this shift and why the work of purification has been abandoned. To offer the requisite conjectures goes beyond the scope of this chapter. I suggested earlier that such work of purification is no longer possible and no longer required today. Some of the preceding remarks offer hints why this is so. For example, improved methods for surveying and managing information from highly complex systems have advanced modeling techniques that are opaque and intractable to the human mind. They do not afford analytic understanding but perfectly robust technical and intellectual control of the phenomena. I have also suggested that the various dimensions of the epochal break thesis require a concerted philosophy of technoscience. Just like the philosophy of pure science, this philosophy of impure science is concerned with questions of epistemology, method, ontology, and representation. It seeks to show how the various characterizations of technoscience complement one another—the epistemological characterization according to which technoscience consists in the acquisition and demonstration of basic capabilities of visualization, intervention, and control; the ontological characterization in terms of objects and processes that can be assigned neither to nature nor to culture; and the historical characterization in terms of an ontological indifference that could not be fully appreciated as long as the sciences were held to the ideal of purification. Work to more fully articulate such a philosophy of technoscience has begun (e.g., Bensaude-Vincent 2009; Echeverria 2003; Nordmann 2008, 2010).

At the end of this chapter, the urgent plea to acknowledge an age of technoscience is surrounded by an air of paradox. It posits an epochal break between a modern age of scientific revolutions or successive worldviews, and a postmodern age of technoscience that has no historical self-understanding but regards all research at all times as knowing by doing, as a means to create and realize technical potential and thus to construct the world we live in. The very diagnosis of *this* epochal break is therefore a reaction against it. Seeking in a very modernist way to give voice to the spirit of my age and to answer the call of the day, I am reacting also against those historians and philosophers of science who have unwittingly embraced a technoscientific attitude by seeing nothing unusual anywhere and only the endless toil of fitting bits of theory to bits of reality. What would awaken these philosophers from their slumber is the claim that technoscientific knowledge is inferior to that of the sciences, that our theories and our technologies are becoming unreliable.

This is not, however, what I mean when I say that the significance of the epochal break can hardly be overstated. On the contrary, though nearly as opaque and complex as reality itself, technoscientific knowledge may well be more robust and reliable than that of the traditional sciences. The significance rests in the dissolution of the alliance between science and Enlightenment and thus in the loss of an institution for organized scepticism. Even if this institution rarely lived up to its promise, in the age of technoscience that promise itself has lost its meaning.

NOTES

1. Not without a sense of irony, Blumenberg (1976) begins his reflections with this reference to Goethe.
2. I cannot detail this here. Most open to the accusation might be John Ziman's "post-academic science" (2000). The proponents of "mode-2 research" have also been accused of this, but unjustly so.
3. Regarding these "earlier attempts," the Vienna Circle, Karl Popper, Thomas Kuhn, or Imre Lakatos referred general patterns in the development of science to the question of its historical purpose or mission. Typical names for this mission are "scientific progress," "physicalism," "unified science," or "reductionism."
4. To be sure, it needs to be shown that notions of objectivity in synthetic chemistry, the engineering sciences, and even the science of sociology also rely on attempts to separate immutable and systemic features from contingent interventions, if only

by conditionalizing the latter ("to the extent that there can be objective sociological knowledge, it consists in statements of the form 'in this kind of social system, this would be observed'").

5. Accordingly, a project by Bernadette Bensaude-Vincent, Sacha Loeve, Alfred Nordmann, and Astrid Schwarz (Forthcoming) has set out to explore the "Genesis and Ontology of Technoscientific Objects."

REFERENCES

Bensaude-Vincent, Bernadette. 2009. *Les vertiges de la technoscience: Façonner le monde atome par atome.* Paris: Editions La Découverte.

Bensaude-Vincent, Bernadette, Sacha Loeve, Alfred Nordmann, and Astrid Schwarz. Forthcoming. "Matters of Interest: The Objects of Research in Science and Technoscience." *Journal for General Philosophy of Science.*

Blumenberg, Hans. 1976. *Aspekte der Epochenschwelle.* Frankfurt: Suhrkamp.

Daston, Lorraine, and Peter Galison. 2007. *Objectivity.* New York: Zone Books.

Echeverria, Javier. 2003. *La revolucion tecnocientifica.* Madrid: Fondo de Cultura Economica.

Forman, Paul. 2007. "The Primacy of Science in Modernity, of Technology in Postmodernity, and of Ideology in the History of Technology." *History and Technology* 23: 1–152.

———. 2010. "(Re)cognizing Postmodernity: Helps for Historians—of Science Especially." *Berichte zur Wissenschaftsgeschichte* 33: 157–75.

Galison, Peter. 1997. "Material Culture, Theoretical Culture, and Delocalization." In *Science in the Twentieth Century*, edited by John Krige and Dominique Pestre, 669–82. Amsterdam: Harwood.

———. 2006. "The Pyramid and the Ring." Lecture at the conference of the Gesellschaft für analytische Philosophie (GAP), Berlin. September.

Latour, Bruno. 1993. *We Have Never Been Modern.* Cambridge: Harvard University Press.

Nordmann, Alfred. 2006. "Collapse of Distance: Epistemic Strategies of Science and Technoscience." *Danish Yearbook of Philosophy* 41: 7–34.

———. 2008. "Philosophy of Nanotechnoscience." In *Nanotechnology: Principles and Fundamentals*, edited by Günter Schmid, 217–44. Weinheim, Germany: Wiley.

———. 2010. "Science in the Context of Technology." In *Science in the Context of Application*, edited by Martin Carrier and Alfred Nordmann, 467–82. Dordrecht: Springer.

Weber, Max. 1919 [1946]. "Wissenschaft als Beruf." English translation in *From Max Weber: Essays in Sociology*, edited by H. H. Gerth and C. W. Mills, 129–56. New York: Oxford University Press.

Ziman, John. 2000. *Real Science: What It Is and What It Means.* Cambridge: Cambridge University Press.

We Are Not Witnesses to a New Scientific Revolution

GREGOR SCHIEMANN

Do the changes that have taken place in the structures and methods of the production of scientific knowledge and in our understanding of science over the past fifty years justify speaking of an epochal break in the development of science? Some philosophical and sociological descriptions of these changes do indeed assert that such an epochal break is becoming apparent (see Forman 2007; Funtowicz and Ravetz 1993, 2001; Gibbons et al. 1994; Nowotny, Scott, and Gibbons 2001, 2003; Ziman 2000; and others). In general, this thesis is formulated in such a way as to compare the extent of the changes that have occurred or that are to be expected to the early modern scientific revolution. The point of departure for the current epochal break is presented as a tradition that persisted from the beginnings of the early modern era until about fifty years ago, and from which recent developments constitute a fundamental departure. With the completion of this transformation, science supposedly will have freed itself from its early modern origins and undergone a second scientific revolution.

I, however, do not concur with this thesis of an epochal break in the development of science (hereafter referred to as "the epochal break thesis" for short). The critical appraisal that I offer has three parts.

1. The Extent of the Current Process of Transformation

I acknowledge that there is a serious basis to the epochal break thesis. It rests upon verifiable and to some extent profound changes in the production of scientific knowledge and in our understanding of science that have been occurring globally, but especially in the industrially more developed countries in recent years. Some of these transformative processes have been the subject of intensive academic discussion and public debate in the past several years. To name just a few focal points of the changes at issue: the scientification of more and more areas of society, the accelerated increase in the prominence of technology and economics within science, the growth in the complexity of scientific objects, and the dissolution of disciplinary structures in certain innovative fields of research. Although some of these changes are of a gradual nature, others are indeed drastic. On the whole, there are enough phenomena to point to for it to appear justified to speak of an epochal change in the development of science.

Of course, this viewpoint presupposes that such changes (or breaks, if they are discontinuous) are possible in the first place, and also that they are observable at the time when they are occurring. As for the first of these presuppositions, I demonstrate in the following section that the transition from medieval to early modern science can be interpreted as an epochal change. However, it is questionable whether the participants in such an epochal process of upheaval can themselves recognize the significance of this process, since they lack the necessary distance to perceive the overall context in which it is occurring. This objection cannot be wholly rebutted. Although we can assume fictional standpoints external to the contemporary world, we always remain involved in the events of our own time. Nevertheless, given that we are creatures that must construct our own histories, we have no alternative. We are compelled to compare the contemporary world with past eras to gain the historical orientation that is indispensable for shaping the present in a reasonable way.

But that is just the beginning of the real difficulties involved in the evaluation of the historical dimension of the present development of science. What are the "verifiable" alterations in the production of scientific knowledge? How can an "understanding of science" be pinpointed? What role can philosophy of science play in this? Is it possible to speak of "science" in the singular? To answer questions such as these, it is necessary to set up suitable criteria and to conduct the right kind of inquiries, and that is what I work toward in this chapter. Some of the criteria refer to historical material, and others to philosophical and sociological analyses of contemporary science.

2. Toward a Historical Location of Science's Departure from the Early Modern Era

The second part of my critique deals with the ahistorical character of the epochal break thesis. It is not plausible to present science as having constituted mainly one type (e.g., mode-1, normal science, academic science, modern science) from the beginnings of the early modern era until about fifty years ago, and then to contrast it with a supposedly new type (e.g., mode-2, postnormal science, postacademic science, postmodern science). In describing the period from the beginnings of the early modern era until the present, it is more appropriate to apply a two-phase model that incorporates criteria according to which it is possible to recognize a shift in the history of science during the course of the nineteenth century.

I begin by using a variant of the epochal break thesis to demonstrate that the postulated break is in fact understood as a departure from the early modern era. Then I weigh the merits of the two-phase model as an alternative. Finally, I use it to critique the criteria set forth on behalf of the epochal break thesis.

2.1. A Variant of the Epochal Break Thesis

The variant I discuss—namely, the shift from mode-1 to mode-2 science described by Michael Gibbons and his coauthors—is probably the most well-known formulation of the thesis (Gibbons et al. 1994; Nowotny, Scott, and Gibbons 2001, 2003). They decisively classify science in the early modern era as mode-1 (Gibbons et al. 1994, 167). The characterization of mode-1 science—aka the "Newtonian model of science"—proceeds by means of five points of contrast with the currently emerging mode-2 (Gibbons et al. 1994, vii, 3, 10).

Mode-2 science, according to the thesis, developed out of mode-1 science only after the Second World War as a result of the drastic increase in the number of trained scientists and in the technical possibilities for knowledge production (Gibbons et al. 1994, 10, 17, and 44). Some of the preconditions, such as the development of extra-academic research and the dissolution of traditional validity claims, can be traced back to the end of the nineteenth century (e.g., Gibbons et al. 1994, 22; Nowotny, Scott, and Gibbons 2001, 197). Today, mode-2 constitutes a distinct form of knowledge production that is "different from mode 1 . . . in nearly every respect" (Gibbons et al. 1994, vii) and that interacts with mode-1 (Gibbons et al. 1994, 14 and 9). In the future, however, mode-1 will be "incorporated within the larger system" of mode-2 (Gibbons et al. 1994, 154). Not until then will the revolutionary break be completed.

2.2. A Two-Phase Model of Early Modern and Modern Science

The model that I would like to juxtapose to the epochal break thesis characterizes its caesura in the nineteenth century as only a partial departure from the
early modern origins of science. A partial transformation might also have an
epochal character. To some extent, the model follows along the lines of other
inquiries (e.g., Bachelard 1938; Diemer 1968; Foucault 1970; Lepenies 1976;
and Schnädelbach 1983).

2.2.1. The Early Modern Phase

The first ("early modern") phase is characterized by its distinctness from medieval science. In contrast to medieval science, science in the early modern era
distinguished between Christian belief and knowledge, introduced the autonomous person as subject of science, disposed of the ancient conceptual distinction between nature and technology, developed experimentation as a method
for attaining scientific knowledge, and discovered the technical applicability of
scientific knowledge—to name just a few of the significant accomplishments
of science in the early modern phase. In sum, the shift from medieval to early
modern science can be characterized as an epochal change (in my view it was
not discontinuous). Any assertion of a present or future epochal change must
be measured against this caesura in the development of science.

The epochal change that initiated early modern science occurred primarily
within the natural sciences, which subsequently rose to become paradigmatic
for science in general. But the new conceptions of science arising here came
to be applied only partially in other disciplines. On the other hand, the natural sciences continued to be reliant upon traditional and necessary criteria of
scientificality that were valid for other disciplines. The decisive conception that
was taken over from medieval science was the "classical conception of science,"
which can be traced back to the ancient origins of science. On the classical conception, scientific knowledge must be marked by generality (in concepts and
judgments), necessity (of systematic connectedness), and truth (Schnädelbach
1983, 106). The concept of truth took on a key role in this conception. It designated a content on the basis of which the essence of an entity could be determined exclusively, and aimed at a general and solely valid system of knowledge
that comprehended the entire world. The classical conception of science was
and remains effective as an ideal within the sciences and in their public presentation. This can be demonstrated not only historically, but also with reference to
current debates about the conceptions of science. That the epochal break thesis

distances itself from the early modern understanding of science is especially clear when one looks at the criticism it makes of the classical conception: current and future science supposedly should be marked not by generality but by particularity, and not by truth but by uncertainty (Nowotny, Scott, and Gibbons 2001, 4–5, 33–37).

Thus, although the modern scientific revolution affected the entire system of the sciences, it was in a twofold sense only a partial change: it left some central features untouched, and it was initially limited to just one area of science. The restricted character of the change corresponds to its only partially discontinuous progress. The notion that the transition from medieval to modern science marked an "epochal threshold" (Blumenberg 1985) is widely contested among historians.

2.2.2. The Modern Phase

In its criticism of the classical understanding of science, the epochal break thesis refers back to the nineteenth century, at which point the classical conception began to collapse and thereby to usher in the second ("modern") phase. Modern science began to distinguish itself from its precursor by impugning the classical claim upon truth, which did not allow for the possibility of revising scientific knowledge. Within the sciences the discussion was led by such researchers as Carl Gustav Jacob Jacobi, Carl Neumann, Bernhard Riemann, and Hermann von Helmholtz. In the case of Helmholtz, a look at the public lectures that were already famous during his lifetime reveals an increasing tendency to do without the notion of truth as a system of knowledge to be attained at some point in the future. The departure from a classical understanding of truth had the consequence of rendering scientific knowledge hypothetical (Schiemann 2009; Heidelberger and Schiemann 2009). The transformation of scientific knowledge claims was taken up within philosophy most prominently by Friedrich Nietzsche. Moreover, the significance of epistemic characteristics that had dominated the understanding of science in the early modern phase was relativized (Lübbe 1986). At the beginning of the twentieth century, quantum mechanics set an example of how the truth of physical knowledge could be irrelevant in a way that had previously not been suspected.

The transformation of science that began in the nineteenth century did not restrict itself to relativizing epistemic criteria. As further characteristics, I would like to mention the incipient entanglement of scientific and societal development, and the crystallization of scientific communities. The technical applicability of scientific knowledge having already been among the central insights of early modern times, the nineteenth century discovered the comprehensive societal utility of the scientific method and of scientific knowledge. The inverse relevance of applied contexts for scientific development took, among other things, the form of increased state sponsorship of the relevant natural-scientific disciplines (including building up their laboratories) and of applied sciences. The production of scientific knowledge thereby came to be tied to its technical application (e.g., the first and second laws of thermodynamics). With the crystallization of scientific communities, the individual as agent of science withdrew to the background.

It is possible to establish connections among the three characteristics of the second phase. I assume that the increasing interconnectedness of science and other societal areas played a sort of key role. It caused the scientific space for pursuing epistemic questions to shrink. At the risk of simplification, one could say that the real insight of the nineteenth century was that science could be highly useful and applicable even without clearing up epistemological questions. Such issues were in a sense put off indefinitely and have since been threatening to fall into oblivion.

2.2.3. The Early Modern and the Modern Phase

One may ask what the relationship is between the two phases and how science at present relates to them. In response, I would say that the second phase takes on some characteristics of the first and that it includes the present. The two phases differ markedly from each other in their appraisal of epistemic questions. With the critique of the classical understanding of science, transepochal criteria of scientificality come to their limit. But if one focuses on the relationship between science and society, the relationship between the two phases appears in a different light. At first glance, then, the second phase may well appear to be just a continuation of the first. In the same sense Martin Carrier, in his chapter in this collection, understands the modern scientific revolution as a technoscientific project. The early modern scientific revolution was already to some extent a result of a close connection between science and societally anchored technology. It already seemed plausible at the time that the technical production and application of scientific knowledge could in the future yield a thoroughgoing improvement of the conditions of life (e.g., Francis Bacon's New Atlantis). Science, however, was practiced primarily by an elite group upon which other societal forces were not really able to exercise any influence. It was not until the nineteenth century, when applications of scientific knowledge took on reality-shaping dimensions, that institutionalized societal forces began to have a lasting impact on the forms of knowledge of production. Although the establishment of mutual, interactive relations between science and society caused the boundary between the two regions to become far more porous, it did not eliminate this boundary altogether. I will return to this issue in section 3.1.

As for the question whether the transformation in the nineteenth century constitutes an epochal transformation, I would like to leave that open. The process does not appear to have reached its conclusion (see section 3.2). As fundamental as some transformations were vis-à-vis the first phase, it is unclear whether they lend themselves to a uniform new characterization or whether it can persuasively be argued that a uniform account is not possible. Current efforts in philosophy and sociology of science to describe the scientific developments that are presently under way (the epochal break thesis being among them) can be regarded as efforts to conceptualize and sum up the second phase in a uniform manner. The absence of a uniform account could be taken to reveal that the process of transformation that began in the nineteenth century has not yet come to a conclusion (see section 3.2). That a process proceeds gradually over a long period of time need not undermine its status as an epochal change (see section 2.2.1). But there are also contrary tendencies that point to a renaissance of classical conceptions—for example, positivism, which aimed to limit scientific knowledge to observable phenomena; pragmatism, which derived claims to truth from the success of scientific theories; and realism, according to which scientific knowledge gradually approaches truth.

2.3. A Critique of the Criteria Offered by the Epochal Break Thesis

In my view the criteria introduced by advocates of the epochal break thesis are insufficient to establish their overall interpretation. They tend either to go back to the early modern scientific revolution or to subsequent developments before the past half-century, or else not be typical of contemporary science as a whole. I confine my criticism here to discussing two examples of criteria that are cited in defense of the epochal break thesis: one that has been proposed by Gibbons et al. (1994), Nowotny, Scott, and Gibbons (2001, 2003), and another proposed by Alfred Nordmann (2007).

2.3.1. Context of Application

This context comprises "problem-solving and the generation of knowledge organized around a particular application[, and not] merely applied research or development. [It i]ncludes the milieu of interests, institutions and practices which impinge upon a problem to be solved" (Gibbons et al. 1994, 167). I would like to distinguish two primary meanings that are both compatible with these stipulations:

1. The practical-technical context, which is determined by society's expectations and is predicated on the development of specific scientific knowledge and its application. This context was already present in the nineteenth century, when the dynamics of the natural and technical sciences began to become entangled with the transformation of society (see section 2.2.2). In their explication of the concept of a context of application, the authors themselves refer to the establishment of the technical disciplines in the nineteenth century (Gibbons et al. 1994, 4). But these disciplines were supposedly either denied scientific status, or else they completed a transition into nonapplied mode-1 academic sciences (Gibbons et al. 1994, 4). I have two objections to this. The technical disciplines were only temporarily denied scientific status, and only formally—namely, in the refusal to allow doctoral titles to be granted in technical sciences. It is true that there were efforts within the traditional academic disciplines to formulate a conception of science that was divorced from applications, and to use this conception also for the developing technical disciplines. But this phenomenon appeared only as a reaction to the more significant and undeniable increase in the relevance of application also for the traditional academic production of knowledge.

2. The context was already shaped by scientific applications (in the first sense), as discussed by Silvio Funtowicz and Jerry Ravetz (1993, 2001). In this context problems arise that are characterized by a high degree of complexity, by our only partial theoretical grasp of them, and by controversy about knowledge-related evaluations. Their solution is of urgent necessity for society and is connected with high stakes (Funtowicz and Ravetz 1993, 86; and Funtowicz and Ravetz 2001, 19). Funtowicz and Ravetz mention the problem of the environment as a paradigmatic example, to which Nowotny, Scott, and Gibbons also attribute a key role with respect to the societal conditions under which mode-2 science is developing (Nowotny, Scott, and Gibbons 2001, 6–8).

I assume that the context of application in this second sense is limited to specific problems that are decidedly atypical of the majority of objects of scientific inquiry. Nowotny, Scott, and Gibbons assert that the specific problems in this context of application can only be adequately treated with the elements of knowledge production present in mode-2 (e.g., the "strong contextualization"). It is revealing, though, that the authors themselves concede that they do not think that these elements have taken on a decisive function for the entire system of the sciences (see Nowotny, Scott, and Gibbons 2001, 131–42).

2.3.2. The Indistinguishability of Nature and Technology

One of the features that Nordmann (2007, 11) points out in characterizing the new type of science that is currently developing, and which is turning away from the modern "project of science," is the "impossibility to separate out the theoretical representation of nature and the technical intervention in nature." That would, according to Nordmann, be the end of the distinguishability between nature, which is the object of theoretical inquiry, and technology, which is mediated by practice. In order to grasp the scope of this claim, I believe it is necessary to make a distinction between two concepts of technology and two corresponding concepts of nature. The early modern scientific revolution devalued the Aristotelian opposition of nature and technology. Instead of having to disregard human actions, science has since then been able to avail itself of technical constructs. But that does not so much eliminate the distinction between nature and technology as invest it with a new meaning. Technology can, for example, be identified as that which can be traced back to human agency. As I have shown in the case of nanotechnology (Schiemann 2005a), it is often possible to pick out parts of technoscientific objects that are not produced by the actions of technicians. It is true that there are an increasing number of objects for which it is impossible or problematic to distinguish between nature and technology. But we are a long way off from a situation where we would only be able to make such a distinction in exceptional cases.

As an example to illustrate the indistinguishability of nature and technology, Nordmann refers to the so-called OncoMouse, a mouse that is genetically modified such that its susceptibility to breast cancer is significantly increased (Haraway 1997). He describes the "technical production of the mouse" as a "stage on which a purely natural phenomenon shows [not] itself" (Nordmann 2007, 13). Even if the process that does not go back to human agency could be identified, it could no longer be characterized as a phenomenon. If this example really did illustrate a general state of affairs, we would be confronted with a world that is ever more technically reshaped—or, as Werner Heisenberg (1953, 412) put it, "confronted only with ourselves." But I do not think that Nordmann's characterization applies even to the OncoMouse example. Even nonprofessional observers can recognize this mouse as an organism that has been seriously damaged by humans, but that does not owe its existence to human agency. One can assume that the damages have consequences for the entire organism and affect all the vital processes of the animal. But this is simply an expression of the animal's natural holistic constitution, which would be also modified by natural injuries. In short, nature remains present even in high-tech laboratories. Indeed, it arises in the form of this mouse's suffering in such a way that it provokes our pity and is thereby part of the motivation to protest against this case of genetic modification.

3. Two Limits on the Current Process of Transformation

The last part of my critical appraisal concerns two limits on the current process of transformation of science and also on the descriptions of this process. These remarks are of a more general and more speculative character.

3.1. Science and Society

The epochal break thesis attributes to science's successes an inordinate societal and cultural significance. It is indeed correct to say that science's increasing ubiquity is extending into ever more societal sectors, and also that society is in possession of better means to influence science since scientific knowledge is more widely available (Nowotny, Scott, and Gibbons 2001, 215–29). But the authors who emphasize this point overlook the more astounding phenomenon in this context: despite the scientific permeation of society, societal domains retain their obstinacy, and science is thus still confronted with nonscientific knowledge.

There are various reasons for the resulting preservation of the boundary between science and society. On the one hand, science is far from having lost its specificity. Scientific education and research take place predominantly within special institutions. It generally involves concentrating for years on a specific area of inquiry, whereby one attains a competency that cannot even be matched by members of other disciplines, let alone by outsiders who do not have an academic education. On the other hand, I would also like to refer to the loss of cultural significance that, as I mentioned earlier, scientific claims to knowledge have suffered since the nineteenth century. The everyday habits and patterns to which people look for orientation are well able to retain their identity, because they are hardly about to be overthrown by innovative scientific knowledge (as they were by Darwin's theory in the nineteenth century or by modern physics at the beginning of the twentieth century).

Moreover, the black-box character of objects produced by scientific technology plays an important role in maintaining the distant character of science in everyday life (in German, *Lebenswelt*; for more on this concept, see Schiemann 2005b, 89–125). Nowadays, such devices are almost exclusively constructed in a way that one can use them without knowing anything about the way in which they function. Besides, they are designed such that their internal functioning can hardly be damaged even by misusing them. In everyday life we are usually confronted only with the surfaces of modern technological objects. Although ever more aspects of life are directly dependent on the use of scientific technology, and this dependency is increasingly changing our understanding of ourselves, the black-box character of this technology constrains its influence within certain boundaries. There is as little need to take an interest in the scientific knowledge at the basis of a technical device as there is to understand the technical workings of the device.

3.2. The Future of Science

The transformations of science subsequent to the early modern phase need not reflect an irreversible departure from tradition in every respect. They could also constitute a development that takes place within an already existing framework and that even might reveal elements typical of earlier phases of science. So, although science ceased to be primarily epistemically oriented, it might in the future once again become so—especially if we do not regard the transformation in the nineteenth century as an epochal change.

Since society discovered the usefulness of science, science has learned to deploy its world-shaping potential. Nevertheless, there has been increasing pressure on science to develop new methods and new knowledge that do justice to societal needs and demands. Science could still possess an epistemic interest divorced from applications, but may not yet incorporate the requisite measure of routine to satisfy societal requests without suppressing this interest. In the future it could be possible for science to avoid being limited to its utility if its role in applied contexts were to take on a more self-evident character. Furthermore, it is imaginable that societal needs and demands upon applications of science could reach a saturation point. Society, then, could again begin to place a greater value on epistemic issues.

REFERENCES

Bachelard, Gaston. 1938. *La formation de l'esprit scientifique: Contribution à une psychanalyse de la connaissance objective.* Paris: Vrin.
Blumenberg, Hans. 1985. *The Legitimacy of the Modern Age.* Cambridge: MIT Press.
Diemer, Alwin. 1968. "Die Begründung des Wissenschaftscharakters der Wissenschaft im 19. Jahrhundert—Die Wissenschaftstheorie zwischen klassischer und moderner Wissenschaftskonzeption." In *Beiträge zur Entwicklung der Wissenschaftstheorie im 19. Jahrhundert,* edited by Alwin Diemer, 3–62. Meisenheim am Glan, Germany: Hain.

Forman, Paul. 2007. "The Primacy of Science in Modernity, of Technology in Postmodernity, and of Ideology in the History of Technology." *History and Technology* 23: 1–152.

Foucault, Michel. 1970. *The Order of Things: An Archaeology of the Human Sciences.* New York: Pantheon Books.

Funtowicz, Silvio O., and Jerry R. Ravetz. 1993. "The Emergence of Post-Normal Science." In *Science, Politics, and Morality: Scientific Uncertainty and Decision Making,* edited by René von Schomberg, 85–123. Dordrecht: Kluwer.

———. 2001. "Post-Normal Science: Science and Governance under Conditions of Complexity." In *Interdisciplinarity in Technology Assessment: Implementation and Its Chances and Limits,* edited by Michael Decker, 15–24. Berlin: Springer.

Gibbons, Michael, Camille Limoges, Helga Nowotny, Simon Schwartzman, Peter Scott, and Martin Trow. 1994. *The New Production of Knowledge: The Dynamics of Science and Research in Contemporary Societies.* London: Sage.

Haraway, Donna. 1997. *Modest_Witness@Second_Millennium, FemaleMan Meets OncoMouse: Feminism and Technoscience.* New York: Routledge.

Heidelberger, Michael, and Gregor Schiemann, eds. 2009. *The Significance of the Hypothetical in the Natural Sciences.* New York: de Gruyter.

Heisenberg, Werner. 1953. "Das Naturbild der heutigen Physik." In *Gesammelte Werke,* edited by Werner Heisenberg, 398–420. Vol. C I, *Physik und Erkenntnis: 1927–1955.* Munich: Piper.

Lepenies, Wolf. 1976. *Das Ende der Naturgeschichte: Wandel kultureller Selbstverständlichkeiten.* Frankfurt am Main: Suhrkamp.

Lübbe, Hermann. 1986. *Religion nach der Aufklärung.* Graz: Fink.

Nordmann, Alfred. 2007. "A New Mode of Research: Arguing for an Age of Technoscience." Manuscript.

Nowotny, Helga, Peter Scott, and Michael Gibbons. 2001. *Re-Thinking Science: Knowledge and the Public in an Age of Uncertainty.* Cambridge, Mass.: Polity.

———. 2003. "Introduction: Mode 2 Revisited: The New Production of Knowledge." *Minerva* (special issue edited by R. MacLeod) 41: 179–94.

Schiemann, Gregor. 2005a. "Nanotechnology and Nature: On the Criteria of Their Relationship." *Hyle—International Journal for Philosophy of Chemistry* (special issue "Nanotech Challenges") 11: 77–96. Available online at http://www.hyle.org/journal/issues/11-1/schiemann.htm.

———. 2005b. *Natur, Technik, Geist: Kontexte der Natur nach Aristoteles und Descartes in lebensweltlicher und subjektiver Erfahrung.* New York: de Gruyter.

———. 2009. *Herman von Helmholtz's Mechanism: The Loss of Certainty. A Study on the Transition from Classical to Modern Philosophy of Nature.* Dordrecht: Springer.

Schnädelbach, Herbert. 1983. *Philosophie in Deutschland, 1831–1933.* Frankfurt am Main: Suhrkamp.

Ziman, John. 2000. *Real Science: What It Is, and What It Means.* Cambridge: Cambridge University Press.

"Knowledge Is Power," or How to Capture the Relationship between Science and Technoscience

MARTIN CARRIER

IT IS DIFFICULT, IF NOT IMPOSSIBLE, to judge the continuities or ruptures involved in a historical process of which oneself is a part. Historians are aware of the human tendency to view one's own period as a turning point in history. Epochal breaks have been diagnosed galore, which we hardly remember anymore. Take the now almost forgotten "conference on security and cooperation in Europe," which took place from 1973 to 1975 in Helsinki and Geneva and was considered at the time as a major turning point in the history of the second half of the twentieth century. But this event is now completely eclipsed by the changes inaugurated in 1989. However, it is not possible to dismiss all such historical claims indiscriminately; sometimes they are correct. Take Johann Wolfgang von Goethe's sentiment at the cannonade of Valmy in September 1792 that a new historical era had begun. He was right, after all, since the French Republican forces (consisting of volunteers) had proven able to hold out against professional Prussian troops. Goethe correctly recognized that this victory ensured the survival of the revolution and signaled the advent of a new era in world history—that is, an epochal break. Conversely, the famous, if contested, entry of *"rien"* (or "nothing") that Louis XVI, King of France, wrote in his diary on July 14, 1789, referring to his lack of luck in the day's hunting, demonstrates that we may overlook grand-scale ruptures happening right in front of our eyes.

1. Three Claims Concerning Recent Breaks of Science with Its Past

It needs to be stressed at the outset that all judgments about a contemporary historical break are highly uncertain and dubious. In matters of this sort, we run the risk of looking like fools in hindsight wisdom. Against the background of this *captatio benevolentiae* I wish to advance three claims regarding the changes science is presently undergoing:

1. The regime of technoscience that Alfred Nordmann views as a recent prod-uct of the postmodern "abandonment of the work of purification" (see his chapter in this collection) is in fact part of the scientific revolution. The seventeenth-century pioneers aimed for knowledge in the service of utility. In the modern era a scientific enterprise in the sense of a project of pure knowledge gain, untainted by practical objectives, has never existed. This observation also serves to undermine Gregor Schiemann's attempt to shift the alleged break, or what comes close to it, to the nineteenth century (see Schiemann's chapter in this edited volume).

2. A different sort of recent epochal break enjoys much more plausibility at first sight. It concerns the replacement of scientific understanding by tech-nological skills. There are some indications to this effect, but this project has proven not viable and will no doubt come to a lusterless ending soon. My estimate is that this attempted reorientation of science will be unsuc-cessful and fail to make it to a break for this reason.

3. At the level of ontology, a technoscientific turn has transformed the objects of scientific scrutiny considerably. Recent natural science seldom addresses entities and processes that exist independently of human intervention. Rather, we mostly inquire into the effects and side-effects of products of our own hands.

All three contentions deal with different versions of the technoscientific project. I surmise that there is continuity as regards the first item, an aborted break as to the second, and an important change (that is, something akin to a break) with respect to the third.

2. The Scientific Revolution as a Technoscientific Project

The first suggestion to heed is Nordmann's insistence that we need to assume a vantage point that supplies us with observations of the appropriate grain. That is, a too fine-grained account and a too coarse-grained description will equally miss the characteristics presently at issue. On the one hand, science has always embraced projects of a divergent character, while on the other hand science has always aimed to construct empirically adequate models. In his chapter Nordmann is quite right in demanding an appropriate "middle term." The corresponding basic distinction concerns the "scientific enterprise" associated with the quest for "truth," or the "determination" of reality, and with "revealing" nature—or with "understanding," the term I would prefer. This scientific enterprise is separated from technology; science turns on "representation," on uncovering nature's machinery.

The "regime of technoscience," by contrast, proceeds in an "engineering mode." What matters is capabilities or skills—that is, the "creation of robust technical systems." The technoscientific project in the methodological sense relevant here is characterized by the notion that theory is subservient to technology or that understanding nature is taken in the service of intervening in nature. I deny that such a historical break has occurred recently. My contrary claim is that the scientific enterprise, characterized in the way sketched, is a historical myth. The scientific enterprise and the regime of technoscience came into the world united; they were born as twins.

More precisely, the scientific enterprise always had a technoscientific commitment at its core. The twin nature of understanding and intervening was constitutive of the scientific enterprise right from the start. It was not the search for truth that marked the pivot of the scientific revolution. Rather, the new idea was to use truths about natural processes as a basis for developing capabilities or skills. Francis Bacon declared that the aim of the new science was to enrich human life and to ease its hardship by new inventions, and the new means recommended to promoting this end was knowledge of the causes and the laws of nature (Bacon 1620 [1990], vol. 1, sec. 81 and 129). The careful and systematic study of nature would eventually bear technological fruit: "But in the true course of experience, and in carrying it on to the effecting of new works, the divine wisdom and order must be our pattern. Now God on the first day of creation created light only, giving to that work an entire day, in which no material substance was created. So must we likewise from experience of every kind first endeavor to discover true causes and axioms; and seek for experiments of Light, not for experiments of Fruit. For axioms rightly discovered and established supply practice with its instruments, not one by one, but in clusters, and draw after them trains and troops of works" (Bacon 1620 [1863], vol. 1, sec. 70). The search for truth, rightly conceived, is conducive to increasing utility.

In the same vein, René Descartes's chief objection to the erudition of the past was that it was barren and had failed to bear practical fruit. The principles of the new science, by contrast, promised to afford "knowledge highly useful in life." If we come to know the forces of nature "as distinctly as we know the various crafts of our artisans, we might apply them in the same way to all the uses to which they are apt, and thus render ourselves the lords and possessors of nature" (Descartes 1637 [1960], chap. 4, sec. 2, p. 101). In sum, the scientific revolution was fueled by the prospect of technological progress. Fundamental research, or knowledge of the laws of nature, is the royal road toward the betterment of the human condition (Bacon 1620 [1990], vol. 1, sec. 110, 117, and 129).

By contrast, pure contemplation had been the hallmark of "science" (or academic knowledge) in Antiquity and the Middle Ages, and it was this kind of scholarship that the scientific revolution had intended to overthrow. Medieval Aristotelianism had separated technology from science in assuming that technical devices distorted the course of nature. Consequently, studying natural processes was considered useless for technological intervention. Conversely, producing a change in these processes was assumed to interfere with their natural course and to impede their understanding. Experimentation as a means of gaining knowledge about nature broke the confines of this conceptual framework. The intertwinement of knowing and doing, as it becomes manifest in experimentation, testifies to the technoscientific orientation in methodology that characterized the scientific revolution.

The scientific enterprise is distinguished at the appropriate grain of description, by connecting truth and utility, or even by taking scientific knowledge as a means for achieving a betterment of the human condition. After all, this is what Bacon meant by his widely received slogan that "knowledge is power" (Bacon 1620 [1990], vol. 1, sec. 129, p. 65). The capacity of intervening in nature needs to be based on knowledge about nature. Science is necessary for building up technology. This entails that science has pursued the track deplored by Nordmann since its inception. The technoscientific break in the methodological respect relevant here occurred directly in the midst of the scientific revolution. Regarding the development of science proper, the technoscientific attitude in methodology, rather, is a historical invariant. We have never been unskilled representationalists.

It is this retained methodological framework that also makes Schiemann's attempt (in his chapter in this collection) to identify something akin to a technoscientific break in the nineteenth century appear dubious. Schiemann claims that in this period science became entangled with technology and acquired a more profound impact on society. I agree with this judgment but hasten to add that this only means that the vision of an applied science or a science-based technology, as conceived around 1600, had gradually become reality. The conception of a science being based on experimentation and mathematical analysis—searching for laws of nature, being capable of supporting technological innovations, and bringing forth social progress—is clearly adumbrated in the writings of Galileo, Bacon, and Descartes. This conception is part of the creed of the scientific revolution.

The search for understanding or "representation" without regard to intervention or skills is mostly part of the ideology of science. Science initially failed to live up to its technoscientific aspirations and this is why the misleading impression was conveyed, or perhaps created so as to veil the failure that science was striving for knowledge for its own sake. In fact, for a long time the fruits did not keep the promises of the flowers; high-brow theory failed to get a grip on the machines and devices. What has been accomplished in the nineteenth century and afterward is the gradual realization of these aspirations. After two hundred years of incessant trial, science finally became sophisticated enough to capture the more subtle features of experience and to cope with the complexity of the phenomena to a degree required for addressing practical challenges successfully. Applied research is particularly demanding in this respect since it cannot confine itself to areas in which effects appear without distortions, idealizations hold, and approximations work satisfactorily. The reason is that in order to be put to technological use reliably, phenomena and effects need to be known and controlled not only generically but also in their more remote properties and intricate aspects. Distortions and side-effects are important in practical contexts; they cannot be neglected by focusing on the pure case—as it is often possible in epistemic science.

Mastering challenges in the context of application bespeaks the amount of progress that has been made in science during the past centuries on the one hand, but also generates new forms of pressure on science on the other. In the past half-century or more, science has been driven to a high extent by practical problems and consists in large measure of mission-oriented research projects. The lion's share of research today is conducted with the stated aim of improving the control of phenomena and to make more advanced intervention possible. Science is seen as a driving force of the economy. All this was different until some 150 years ago, but this change of attitude is merely tantamount to a fulfillment of the original self-understanding of the scientific enterprise.

As a result, there have been important changes in the relationship between science and technology. However, the goal of producing knowledge for use and the idea of supporting intervention by understanding—that is, the perceived means-end relation of *scientia ancilla technologiae*—have remained stable over the centuries. Rather, scientific progress has brought us into a position in which science could eventually live up to the expectations of the pioneers of the scientific revolution. That science and technology actually are the tandem enterprises they were intended to be in the seventeenth century is due to the success of science. Eventually reaching a goal aspired after for centuries is a far cry from a rupture.

3. Replacing Understanding with Control: An Aborted Technoscientific Break

In fact, a different interpretation of a technoscientific break in a methodological sense in the recent past enjoys much more credibility at first sight—but equally fails eventually. The prima facie more plausible interpretation of a relevant rupture concerns attempts to bypass understanding and to replace representation with intervention. Technoscience in this sense is solely committed to the goals of intervention and control, and no longer to the objective of truth, its attenuated proxies, or understanding. The evidence supporting this claim is furnished by the realization that the relationship between scientific understanding and technological intervention is complex, subtle, and tenuous in that only the conceptual structure of the local models used for coping with the phenomena is typically shaped by theory, whereas the nitty-gritty has to be read off from the data. Model building cannot proceed top-down but needs to include generalizations and specific information without theoretical backing. Applying scientific knowledge does not issue in a pyramid-like propositional structure but rather in a patchwork of related local models (Carrier 2004b).

Worse yet, in some cases new technological skills were not based on scientific representation at all. It took some seventy years until the mechanism of how acetylsalicylic acid (or aspirin) managed to relieve headache was unveiled. The Nobel Prize for chemistry in 2007 was awarded for accounting for the mechanism underlying the Haber-Bosch process of ammonia synthesis, which was introduced at an industrial scale in the 1910s. It follows, then, that the relationship between truth and utility turns out to be not always as close as anticipated in the scientific revolution (Carrier 2007). It is essential to take all allegations of supposed breaks with the scientific past as referring to the practice of science, not merely to its ideology. I grant at once that lots of the more visionary proclamations associated with nanoresearch, for example, are in conformity with a break thesis of this sort. Yet the fact that ideas figure in popular books about the future of science and technology does not suffice as credentials for being adequate descriptions of science.

Indeed, what we sometimes find is a rhetorical severing of conceptual ties between science and technology, which is moreover occasionally borne out by historical facts. Examples of the former constitute the visions of self-proclaimed prophets of nanoscience; examples of the latter are furnished by cases like the ones mentioned. And such cases could in principle buttress a technoscientific epochal break thesis—if of a different kind than the one entertained by Nordmann. However, my claim is that cases of this sort are still the exception rather than the rule. The more common relationship between science and technology is the one presented earlier—namely, that theory goes some way toward capturing the relevant data but needs to be supplemented by details read off from experience. I take the following example of a failed declaration of independence of gene technology from biological understanding as characteristic: it illuminates bold claims advocated in this field—and their collapse a few years later.

Genetic engineering frequently draws upon "contextualized causal relations." Such relations are confined to "typical" or "normal" conditions and leave the pertinent causal processes out of consideration—that is, they are not based on scientific understanding. The way the so-called eyeless gene is put to use is a case in point. Eyeless controls for the eye morphogenesis of Drosophila and other organisms like mice. If the operation of the gene is blocked or lost, no eyes are formed—which is why the gene is somewhat misleadingly called "eyeless." If the homologous mouse gene is implanted and expressed in Drosophila, it instigates the formation of fly eyes, not mouse eyes.

The expression of the eyeless gene in suitable tissue is sufficient for eye formation—that is, eyes can be generated by appropriate stimulation in the legs or wings of flies. This is the reason for calling eyeless the "master control gene for eye morphogenesis." But the eyeless gene only sets off a complex series of intertwined genetic processes that only in their entirety control eye formation. This is evidenced by the mentioned fact that the homologous mouse gene stimulates the expression of fly eyes in fly tissue. The eyeless gene operates as a trigger that needs the appropriate causal environment to become effective (Fox Keller 2000, 96–97). The upshot is that the identification of eyeless genes allows the control of eye morphogenesis without theoretical understanding of the underlying processes.

In the 1990s results of this sort were taken as a basis for a declaration of technology's independence from science. The background to this judgment of irrelevance was the scientific assessment that cell activities in general could not be understood by drawing on genetic factors alone. Rather, protein interaction was increasingly considered essential to accounting for cellular operation. However, as the example of the eyeless gene amply testified, no resort to understanding was necessary to control the relevant processes. Contextualized causal relations were sufficient for providing a fixed connection between genes and organismic features or between genes and proteins, on whose grounds targeted interventions in organismic processes looked feasible. Genetic manipulation was considered suitable for producing effects in a uniform, predictable fashion, albeit the reliability of the relevant causal relations is constrained to particular conditions (Fox Keller 2000, 141–42). Biotechnologists of the period argued that the assumption of a close connection between gene and organismic property was scientifically dubious, to be sure, but still provides a handle for intervening in cellular processes. Genes are tools for bringing about intended effects, and this is what biotechnology is all about: identifying levers to pull and switches to press. Taking the underlying causal processes into account tends to alienate biotechnology from of its proper purpose (Bains 1997). The common ground of these judgments is that biotechnology rests on contextualized causal relations whose appropriateness is independent, in large measure, from the truth of more fundamental sophisticated theories. In particular, successful intervention need not rely on disclosing the relevant causal mechanisms. Knowledge and control are taken to be decoupled, and the commitment to truth is sacrificed for the capacity of intervention (Carrier 2004a).

Yet such statements are far removed from present-day reality. Specifically, such declarations to the contrary notwithstanding, we are witnessing a revolution in biotechnology that amounts to the inclusion of proteomics and the partial supplant of genomics by proteomics. The control of gene expression has gained prime importance for biotechnological research endeavors. Genes need to be switched on and off, and this is achieved by the action of proteins. Such proteins are in turn produced by other genes in the cell or stimulated by other influences from within the cell or from outside. The activity of a given gene is promoted or inhibited by a plethora of genetic and nongenetic factors and thus depends heavily on its context. In stark contrast to the eyeless gene, "distalless" acts in a more specific way and affects embryonic development differently. In caterpillar embryos the expression of distalless induces the formation of legs, whereas in the developed butterflies the same gene generates colored eye-spot patterns on the wings (Nijhout 2003, 91). Obviously, in some instances the context is of critical significance for intervening reliably.

Similarly, persons often differ in their precise genetic makeup and the nature of the associated regulatory network. As a result, manipulating one gene may have different effects in different people. This is why genes are not simply universal switches to press; what they accomplish is dependent upon the particular physiological context. Consequently, if one wants to produce effects dependably, the reticulation of influences should better be known.

The low probabilities with which diseases can be attributed to specific genetic causes testifies to this complexity of interaction. According to present estimates, only up to 10 percent of the incidences of breast cancer can be traced back to genetic deviations (i.e., mutations in the BRAC1 or BRAC2 gene). Likewise, only 3 percent of the incidences of Alzheimer's disease can be accounted for by genetic disposition and malfunction. The remainder apparently is produced by interactions among proteins that transcend direct genetic control. The claim that technological skills had been severed from theoretical penetration has thus proven premature and mistaken in this case. The initial declarations of the primacy of a pragmatic, intervention-centered approach have turned out to be self-defeating. Renouncing the quest for understanding has led to technological failure.

To assess the nature of the regime of technoscience, we should not give heed to the general declarations and visionary proclamations. We should rather take notice of whether the bold claims proved viable. As the earlier example demonstrates, technoscientific reality may be a far cry from the expectations of some protagonists of technoscience. Regarding the actual practice, the relationship between science and technology today agrees roughly with what it has been like for the past 150 years—and it is rather in conformity now with what the pioneers of the scientific revolution had envisaged than in the two centuries following this revolution. Actually, the dependence of technological development on scientific knowledge is even more pronounced today than it has been earlier because scientific theory has tightened its grip on the phenomena considerably.

Consequently, a methodological rupture between understanding and intervening, as announced by some advocates of a technoscientific break in latter-day science, has never materialized. To be sure, there are indications pointing in the direction of such a break, but they have turned out to be of a much more modest impact than anticipated. The breakaway was aborted right after its groping beginning. Granted the usual limitations of the theoretical grip on the phenomena in their complexity, scientific understanding, and the theoretical basis on which it rests, will retain its crucial role in science in the context of application.

4. The Technoscientific Turn in Ontology

The most important reorientation of recent science does not arise at the methodological level but at the ontological one. It concerns the "technoscientific turn" regarding the objects of scientific research. Many objects of scientific study are no longer part of untamed nature; they are produced artificially. The technoscientific turn at the ontological level means that most of the entities and processes addressed by scientific research are not found or uncovered but are rather created. Laser processes, chlorofluorocarbons (CFCs), nanotubes, or layered systems of ferromagnetic materials and chromium in a magnetic field (in which giant magnetoresistance makes its appearance) would have hardly ever come to exist by nature left to its own devices. They are deliberately produced by humans but explored like objects generated by the whims of nature.

The two traditional types of relationship between science and technology are, first, that science is instrumental in producing technology, and, second, that technology is used in experiments for gaining new knowledge about nature. The rise of technoscience means that a third mode has gained ascendance. Technologically produced entities or processes have become the objects of scientific scrutiny. We have made them, but we fail to understand their causal or nomological properties. Such objects are human creations, but they offer surprises just as objects untouched by human hands. They need to be studied scientifically to be understood. I hasten to admit that this technoscientific mode of research is not new. It started perhaps with Sadi Carnot's analysis of the steam engine in the framework of caloric theory (Carnot 1824). What is novel, instead, is that the technoscientific mode has become dominant over the past decades. Sometimes the entities and processes produced are intended to mimic nature—such as artificial vitamins or the particles brought forth in the huge accelerators, some of which have made their only unaided appearance in the first second after the Big Bang. Yet the more usual case is that the objects of scientific study have never been part of untouched nature. Neither liquid crystal displays (LCDs), light emitting diodes (LEDs), optical digital discs (CDs or DVDs), nor nonsteroid anti-inflammatory drugs (NSAIDs) owe their existence to nature's contrivances.

Of course, we may conceptually distinguish between nature and technology by saying that nature is what we have not made (see Schiemann's chapter in this collection as well as Schiemann 2005). But we need to adduce the proviso that nature in this restricted sense does in no way exhaust the scope of the natural sciences. Nature that is subject to scientific research transcends by far the realm of what exists independently of us. The former now includes the emergent properties and unforeseen side-effects of our own creations. Traditionally, nature was often regarded as a creative force; now it is we who create nature. Schiemann refers to Werner Heisenberg's saying, by distancing himself from this sentiment, that "for the first time in history humans are faced only with themselves" (Heisenberg 1984, 412). Yet this is exactly what the technoscientific project entails at the level of ontology: We look into nature but what we see is mostly a reflection of our hands. The world has become a mirror of humankind.

Medieval Aristotelianism was characterized by an opposition of nature and technology, the scientific revolution proclaimed technology to be applied science, the technoscientific turn entails that nature is applied technology. If there is any recent break associated with science in the context of application, it does not so much concern the methodological relationship between representing and intervening or between understanding and shaping nature. Rather, it concerns the technoscientific ontology of nature as a human creation.

REFERENCES

Bacon, Francis. 1620 [1990]. *Neues Organon*, edited by Wolfgang Krohn and translated by Rudolf Hoffmann [Latin-German edition]. Hamburg: Meiner.

———. 1620 [1863]. *The New Organon*, edited and translated by James Spedding, Robert L. Ellis, and Douglas D. Heath. In *The Works VIII*. Boston: Taggard and Thompson.

Bains, William. 1997. "Should We Hire an Epistemologist?" *Nature Biotechnology* 15: 396.

Carnot, Sadi. 1824. *Réflexions sur la puissance motrice du feu*. Paris: Bachelier.

Carrier, Martin. 2004a. "Knowledge and Control: On the Bearing of Epistemic Values in Applied Science." In *Science, Values, and Objectivity*, edited by Peter Machamer and Gereon Wolters, 275–93. Pittsburgh: University of Pittsburgh Press.

———. 2004b. "Knowledge Gain and Practical Use: Models in Pure and Applied Research." In *Laws and Models in Science*, edited by Donald Gillies, 1–17. London: King's College Publications.

———. 2007. "Wissenschaft im Dienst am Kunden: Zum Verhältnis von Verwertungsdruck und Erkenntniserfolg." In *Natur—Technik—Kultur: Philosophie im interdisziplinären Dialog*, edited by Brigitte Falkenburg, 15–55. Paderborn: Mentis.

Descartes, René. 1637 [1960]. *Discours de la méthode*. Hamburg: Meiner.

Fox Keller, Evelyn. 2000. *The Century of the Gene*. Cambridge: Harvard University Press.

Heisenberg, Werner. 1984. *Physik und Erkenntnis, 1927–1955. Gesammelte Werke: Allgemeinverständliche Schriften I*. Munich: Piper.

Nijhout, Frederik H. 2003. "The Importance of Context in Genetics." *American Scientist* 91: 416–23.

Schiemann, Gregor. 2005. "Nanotechnology and Nature: On Two Criteria for Understanding Their Relationship." *Hyle* 11: 77–96.

Climbing the Hill
Seeing (and Not Seeing) Epochal Breaks from Multiple Vantage Points

CYRUS C. M. MODY

ALFRED NORDMANN, IN THIS EDITED VOLUME, lays out several sophisticated and plausible arguments for seeing today's science as undergoing an epochal break. Unlike many epochal break believers, Nordmann recognizes the near-impossibility of convincing epochal break skeptics simply by inundating them with facts about *Diamond v. Chakrabarty* or the Bayh-Dole Act. Instead, he wants to shows skeptics a path to a "vantage point" from which an epoch-making transition from science to technoscience is visible.

The vantage point from which an epochal break is presently visible is not where I usually locate myself, but it is not a great stretch to climb up there. Nordmann writes that one reason many historians and sociologists of science (such as myself) are epochal break skeptics is that they are stuck down in the valley of microstudies, where the details of science are in continuous flux, but general trends are invisible. Only by climbing up the hill, and viewing those microstudies from a distance, does an epochal break become visible. Taking a broader view of contemporary public discourse about science, it does look obvious that pronouncements of a new way of doing science have become rather commonplace. Some technoscientific activities, such as the patenting of academic research, clearly have become more routine than they were in 1970. And it seems obvious that some relatively recent technologies, such as personal computers and the Internet, have had a pointed effect on the conduct and values of science. So far, so good.

But a vantage point isn't necessarily a stopping point. If we've gone to the trouble of climbing this high, shouldn't we see if it's possible to climb further? The present epochal break is said to be a fracturing, over the past three decades, of the centuries-old partnership of science and Enlightenment. If, however, we examine that partnership not from a vantage point where we can view the past thirty years, but from a vantage point where we can see the past few centuries, what will we glimpse? In the next section I describe some of the historical landscape that a higher vantage point reveals—a landscape littered with epochal breaks, each bearing a family resemblance. The recurrence of such breaks is, of course, known to those who argue that the current one not only exists, but is more far-reaching and important than its predecessors. Nordmann, for instance, argues that the perpetual proclamation of epochal breaks is a defining feature of modernity, so it is no surprise that a higher vantage reveals more of them. He also argues that the desire to keep climbing to the vantage point at which the historical landscape becomes monotonous and the current epochal break appears trivial is itself a product of the current epochal break.

This is also the argument of Paul Forman (2007), another epochal break diagnostician, in his monumental lament for the death of science at the hands of technoscience. For Forman the fact that most historians of technology do not view the current break (if they concede that it exists) as singularly important indicates just how singularly important (and crippling, and pervasive) it is. For Forman, not seeing the enormity (both in the sense of size and monstrosity) of the epochal break amounts to complicity in the overthrow of the Enlightenment. What I want to do in this chapter is to take the opposite tack. I will argue that from some vantage points a *deflationary* view of the present epochal break is empirically coherent and not necessarily equivalent to a *complacent* view.

I.

One reason Forman laments (and Nordmann urges caution) is that the current epochal break is allegedly shifting science away from its "commit[ment] to an unending quest for truth" (see Nordmann's chapter in this edited volume). On this view the pursuit of truth is perhaps not expelled from the new science, but it is now seen as secondary, or as inextricable from the production of socially useful (or profitable) applications. For Forman the era of science was characterized by an understanding that the means justified the ends—that the methods that ensured that science was uninterested in applications were what made the knowledge it produced worthwhile. The new era, he says, is characterized by an understanding that the ends justify the means—that so long as socially useful technologies result, whatever technoscientists do is acceptable.

Now, announcements that scientists are no longer interested in knowledge as an end rather than a means are as old as science itself. And such complaints have often been tied to proclamations that science has entered a new epoch or that the generation of new knowledge has become foreshortened by (or embedded in) the application of that knowledge. Take, for instance, the postwar scientific disciplines in the United States, beginning with physics. From 1945 to 1951 the number of physics PhDs being churned out by American universities doubled every 1.7 years (Kaiser 2002). Graduation rates then leveled off for a while, but after *Sputnik* they shot up again, doubling every 6.2 years from 1958 to 1968. Federal funding for basic physics research went up by at least a factor of twenty from 1938 to 1953. Physics experienced this inflation first, but practically all scientific disciplines followed suit. As a result, the leading institutions of scientific research ballooned almost without bound. Stanford University, for instance, administered $127,599 in government contracts in 1946; a decade later $4.5 million; a decade after that, $13.5 million (Leslie 1987).

If anything counts as an epochal break, surely this demographic and fiscal shock should. The spike in money and enrollments changed the methods of American physics by incentivizing use of fast calculation techniques, such as Feynman diagrams, in which students could be quickly trained, rather than discussion of fundamentals requiring prolonged student-teacher interaction. The money bubble also influenced the *content* of American physics. Such fields as solid-state and nuclear physics, in which students could be trained quickly and would have jobs waiting for them, grew rapidly. David Kaiser (2004) has argued that phenomena such as gravitation, which were conceptually thorny and yielded few applications, drew proportionally much less study than before World War II.

Many older physicists complained that the pedagogical style required by the rapid expansion of their discipline produced a new generation who had no individual creativity, who had "skills" but little "wisdom" or physical intuition, who were judged solely for their abilities as "team players" rather than as truth-seekers (Kaiser 2004). They also complained about the students' values—that they were "clock-punchers" who saw research as a job rather than a calling, who merely wanted a comfortable life rather than the pleasure of seeking truth. Students, meanwhile, identified the same trends but saw them in an entirely different moral light. Many embraced the notion that physics was a track to middle-class comfort and security. A few senior physicists who oversaw the Cold War's trademark large research organizations—such as Luis Alvarez or Ernest Lawrence—were likewise happy that the new pedagogical regime could supply them with team players able to submerge their egos and live within a bureaucracy. These researcher-managers expressly saw Cold War security needs as requiring an experimental style (and a new pedagogy) that mixed up ends and means in exactly the way Forman sees as characteristic of *today's* epochal break.

In particular, they saw nuclear weapons as experiments at least as much as they saw them as bombs. The great fiscal and demographic shock was, after all, predicated on a view that science should (and could) contribute to the nuclear state. Whole disciplines were reorganized to make this possible: geodesists studied the Figure of the Earth to make missiles fly straight; seismologists found ways to verify test-ban treaties; biologists examined the effects of radiation; linguists and anthropologists devised ways to tell our descendants to stay away from nuclear waste. Technoscience in the Cold War mode was not limited to one or two precocious individuals, such as Alvarez or Lawrence; it was the framework that redefined practically all science and engineering disciplines in the United States—and in much of the rest of the world.

The paradigmatic Cold War technoscientists were the chemical engineers, metallurgists, physicists, mathematicians, and others who researched new nuclear "devices." Those devices weren't just envisioned as the eventual applications of postwar science—they were seen as the draw, and the means, to *do* science. As John Wheeler said of Princeton University's physics graduate students in 1951, "It will be hard for them to do better than [to work] on the thermonuclear project for all-around range of ideas" (quoted in Galison and Bernstein 1989, 320). "Many physicists were entranced by the prospect of creating a phenomenon on earth—for the first time—that before then had only existed in the heavens: the chain fusion reaction" (Galison and Bernstein 1989, 276). Norris Bradbury, the first postwar director of Los Alamos, declared in 1945 that testing such a weapon would "provide some intellectual stimulus [and] might even be FUN" (Galison and Bernstein 1989, 277–78). The whole infrastructure of nuclear weapons development was designed to coproduce truth and bombs—from the by-products of plutonium breeder reactors that became an indispensable part of biological and ecological research to the giant nuclear craters in Nevada that provided data for geologists and climatologists.

It's hard for me to see much difference, then, between the nuclear bomb and objects such as the OncoMouse that are continually cited as evidence of a new era of technoscience. What to make, for instance, of Project Plowshare, the

Atomic Energy Commission's attempt to use nuclear explosions to carve out artificial harbors and canals (Kirsch 2005)? Edward Teller and other proponents consistently described Plowshare as an "experiment"—not just in the sense of something untried, but in the sense of a means for generating knowledge in soil science, climatology, geology, and physics. Yet, not least because of secrecy concerns, there was virtually no attempt to generalize that knowledge to anything other than the experimental system (bomb + soil + air) itself.

Similarly, what about the physicists and mathematicians who developed game theory in the 1950s, partly through simulations of political crisis and nuclear warfare? No one at that point (or fortunately since) had ever fought an extended nuclear war, so these physicists and mathematicians were able to assert that their technical expertise trumped the combat experience of military officers (Ghamari-Tabrizi 2000). The outcome of that expertise was the political war game—a protracted, multiperson, embodied, sometimes highly emotional method for better understanding group dynamics during crisis. Almost all the participants in these games felt they were learning something about group dynamics; yet there was precious little attempt to generalize that knowledge into a sociological theory of groups, or to divorce that knowledge from the national security infrastructure. Nor did most participants even believe that what they had learned was "true" in any objective sense. Rather, what they learned from the simulation was the excitement and emotion and chaos of crisis, a subjective truth that might or might not have any bearing on behavior in a "real" crisis.

II.

So here we have mathematicians and physicists in the 1950s, eagerly developing a new research tool, yet not particularly worried that that tool can't, even in principle, generate objective, enduring truth. I can't see the difference between Cold War tools, such as nuclear earthmoving or political war games, and technoscientific, postepochal break tools, such as the OncoMouse. Nordmann and Forman may mean that there is no difference, except for the gestalt switch of seeing a difference. Nothing can prove a difference between the OncoMouse and the atom bomb, but sitting in 2008 rather than 1958, we have access to a vantage point that lets us announce a difference between them.

But why should the current moment be *the* epochal break, the one that separates modern technoscience from the scientific enterprise that arose in conjunction with modernity and the Enlightenment? Why should this break be any more epochal than the one around 1945—which many people at the time viewed as epoch-making in terms of the conduct, content, values, and point of scientific research? Indeed, many saw the consequences of that epochal break in terms that sound very similar to the way the current epochal break is often described. Whether they viewed the new epoch as a good thing or not, many Cold War scientists believed their peers were becoming more superficial, more interested in experiments that were simultaneously applications, and less interested in knowledge for knowledge's sake.

Why should even 1945 be *the* epochal break, though? Certainly, something important happened then—something at least as epoch-making as any turn in 1980 or 2000—but at second glance we find all the same tropes floating around in 1910 or 1880. Take the case, for instance, of Irving Langmuir, the Nobel laureate chemist who spent most of his career in General Electric's research labs. When Langmuir arrived at GE in 1909, he was tasked with preventing the company's lightbulbs from blackening too quickly (a common problem at the time). Almost the only restriction on his work was that any patentable discoveries would belong to GE. But Langmuir was free to choose any method of arriving at those discoveries—even if that meant he drifted well away from the immediate, applied problem with lightbulbs.

Before the era of big corporate research, such a problem would have been approached largely through trial and error. Various parameters of lightbulb construction would have been tinkered with until blackening became less of a problem, and empirical observations of which variations worked best would have been folded into paths for further trial and error. Langmuir's insight, based in his training as a physical chemist, was to treat the lightbulb not solely as a product in need of fixing, but as an experimental apparatus capable of generating extreme conditions and thereby revealing new data on the nature of matter. In the end he used GE's lightbulbs to develop a theory of adsorption and desorption of ions that won him the Nobel Prize. But, as George Wise (1983, 19) puts it: "Langmuir's descriptions of these matters were not just outstanding science. They also taught GE how to build more efficient lightbulbs, and even helped put the company into the new business of electronics."

Of course, Langmuir's Nobel Prize was awarded not for his improvement of GE's lightbulbs, but for the generalized theory of matter derived from them. Indeed, Langmuir was ambitious to be taken seriously by his academic colleagues for something other than lightbulbs. Yet it is worth pointing out that in his more philosophical moments Langmuir was rather enamored of Percy Bridgman's operationalism—in which, presumably, attempts to take lightbulb knowledge past the lightbulb would be treated with caution. Admittedly, Langmuir rated his journal articles (where he advanced more generalized theories) as more personally satisfying than his patents, but he saw patenting as indispensable. Moreover, both he and GE's lawyers clearly saw journal articles not simply as contributions to chemists' quest for truth, but as ways to build an intellectual property advantage for the company in disputes over technologies such as the thermionic valve. To return to Forman's formulation, it's not clear that this was in fact a period when the means justified the ends. To all appearances, it was a period when the means and the ends were mutually justifying each other and bootstrapping each other into existence.

Nordmann would respond, I think, that we can always find examples prior to any epochal break of people who exhibit the characteristics of the postbreak era, yet that the nature of the break lies in the *seeing* of a difference in the entire scientific enterprise before and after the break. Maybe so, but in Langmuir's time the scientific enterprise was undergoing wholesale changes that look strikingly similar to descriptions of today's epoch. Langmuir may have been exceptionally brilliant, but he was not an exception in his attitudes toward science "in the context of application." Indeed, as Philip Mirowski and Esther-Mirjam Sent (2007) have shown, the whole academic infrastructure that underwrites a conception of pure science conducted without view to application was co-constructed at the end of the nineteenth century with the infrastructure of *corporate* research. The high-tech companies of the day (Siemens, GE, Westinghouse, and so on) needed the academic world to train people like Langmuir who could move beyond trial and error. And to attract those people, companies were perfectly willing to pay for Nobel-winning research that could never be purified into basic and applied portions.

Indeed, in the early part of the twentieth century, scientists working in private, for-profit institutions easily outnumbered those working in universities. Corporate researchers routinely won Nobel Prizes throughout the past century, before and after the putative epochal break. In such disciplines as condensed matter physics, it was very difficult between, say, 1950 and 1990 to find any leading researcher in the United States (and to some extent in Europe and Japan) who had not spent formative years working for a for-profit lab. Corporate research wasn't all *applied* research, but it also wasn't paid for out of an altruistic search for truth. Long before the putative epochal break, the entanglement of corporate and academic research was generating much that looks "technoscientific"—for example, the laser. Epochal break diagnosticians therefore need to more compellingly explain just what distinguishes today's scientists from these forebears.

So should *the* epochal break be pushed back to 1910? Or can we climb a little higher up the hill to yet another vantage point from which that break looks no more epoch-making than the ones at 1945 and 1980? Well, if we keep climbing, we can certainly find people who had the same "technoscientific" attitude as Langmuir in 1910 or Edward Teller in 1955 or Craig Venter in 2000. Take, for instance, William Thomson, Lord Kelvin. As Norton Wise (1988, 92) has shown, in Kelvin's dispute with the Maxwellians over electromagnetism, Kelvin's main argument was exactly that Maxwell's ideas were *not technoscientific enough*: "Thomson employed the practical reality of the telegraph as at once a moral and an epistemological weapon against what he regarded as the metaphysical ideality of Maxwellian theory." For Thomson the telegraph mediated "between two sets of beliefs about progress: industrial progress based on control of matter and scientific progress based on explanation of matter." Science might be a quest for truth, but it could only get there via practical technologies.

That is, for Thomson, "applied science" was redundant—the order and discipline needed to make science *useful* was a requirement for making science at all. "Neither theory, nor instruments, nor measurements, nor even standard units of measurement existed prior to the ocean telegraph venture" (Wise 1988, 94). Once the telegraph was in place, though, physical understandings based on experience making the telegraph a working technology would necessarily take priority over armchair theories. Moreover, Thomson believed any good theory could only be derived from embodied familiarity with commonplace, *useful* artifacts. For instance, "since experiment required light to consist of transverse waves, and since the only media known to transmit transverse waves were elastic solids, the ether ought to be regarded as like other everyday solids, 'Scotch shoemaker's wax,' pitch, or 'calf's foot jelly'" (Wise 1998, 97). Maxwell's ether, which behaved like no known substance (and certainly no commercial product), was unimaginable and therefore unintelligible.

As Norton Wise (1998, 97) put it: "The analogy thus carried a criterion of commercial success into electromagnetism as a criterion of a valid theory." Note that the criterion was not that an electromagnetic theory be applicable in a successful technology. Rather, a robust electromagnetic theory had to be rooted in the infrastructure and operation of a preexisting commercially successful technology. Kelvin was perhaps an even more hard-core technoscientist than today's academic entrepreneurs. Twenty-first-century biotechnologists only imply that their theories are right because they might someday lead to a profit; Kelvin believed that you had to have a profit already in hand before you had the right to espouse a scientific theory.

So, is it that our current epochal break merely takes us *back* to 1884, or was Thomson just 120 years ahead of his time? Certainly, he was more commercial than many of his colleagues. Yet Thomson was also the most influential member of the British scientific establishment of his day, a man who exerted enormous power in directing resources and creating institutions. His vision of conducting research in large, skill-differentiated, multiproject academic lab groups created the template for British (and, more generally, Anglophone) universities right up to the present.

From these higher vantage points, then, we can see how common those moments are when historical actors saw things changing rapidly to accommodate something very much like "technoscience." We can also see just how many continuities there are across those moments. There might be breaks, but they don't seem very epochal, unless an epoch only lasts a generation or so. That, however, doesn't seem to be the meaning of "epoch" that most believers in the epochal break use. Rather, they see the epoch that is putatively ending as having begun with the Scientific Revolution, and as having reached its mature form in the Enlightenment.

Yet even if we look back at the founding of the scientific enterprise, we see many of the same features that supposedly define the new epoch coming into being today. Take, for instance, Galileo's workshop—surely a central site in the building of the scientific enterprise. Once again, we find the same technoscientific attitude to the means of generating knowledge as we have in other periods. As Mario Biagioli (1993) tells it, the reason Galileo's discoveries could be taken as objectively true was *not* because they could be purified from the patronage that made them possible or from the applications both Galileo and his patrons expected from them. Rather, Galileo's discoveries could be taken as objectively true precisely because he successfully obtained patronage from powerful people (the Medicis) who appreciated the political and technological applications of the ideas that he offered them.

Ironically, Galileo desperately needed Medici patronage just so that he could remove the appearance that his discoveries depended on patronage. Well, no contradiction with the epochal break picture of the scientific enterprise there—Galileo at least wanted to appear to be creating durable, objective knowledge. Except we have to ask—why was Medici patronage necessary to connect experimental activities (e.g., looking at the heavens through a telescope) to the quest for truth? Biagioli explains that Galileo's training and disciplinary affiliations precluded his work from being seen as objective before Medici patronage. As a mathematician and astronomer, Galileo was seen by academic natural philosophers as thoroughly embroiled in the workaday world of surveying, navigation, and astrology.

All those were *commercial* activities, ways of making a living by selling expertise to a customer. Because it is difficult to maintain a lasting relationship with a customer if you don't tell them what they want to hear, the academic natural philosophers believed that no mere mathematician could be trusted—the applied nature of their work prevented it from being objectively true. Even in principle, Galileo's description of the universe couldn't be anything more than a heuristic in solving practical problems for pay. Thus Galileo convinced the Medicis to appoint him as the court *philosopher*, not mathematician—thereby leveraging patronage to lift his discoveries out of the commercial mud and into the realm of objectivity.

Of course, academic natural philosophers opposed that transformation. So, right at the beginning of the scientific enterprise, we see the same rhetoric that characterizes today's talk (from both supporters and critics) about an epochal break in science. On the one hand, the academic natural philosophers (much like Forman) worried that commercially minded expertise was unjustifiably being raised to the status of objective knowledge. And on the other hand, scientific revolutionaries such as Galileo celebrated (much like Mike Roco or Craig Venter today) their breaking down of barriers raised by the traditional disciplines and their forging of a new philosophy that would work hand-in-hand with "industry" and the state. Galileo, after all, ran a workshop that manufactured telescopes, military compasses, and other instruments for the Venetian (later Medici) military and diplomatic corps.

III.

Nordmann closes his chapter by saying that "the very diagnosis of *this* epochal break is therefore a reaction against it, . . . [and also] against those historians and philosophers of science who have unwittingly embraced a technoscientific attitude by seeing nothing unusual." Forman comes to much the same conclusion—that ideology has blinded historians of technology to the primacy of science over technology before 1980. In Forman's eyes, historians of technology have been so busy trying to perceive the primacy of technology over science in the pre-1980 period that they were unable to notice, much less be shocked by, the actual reversal of primacy after 1980. If you can't see the epochal break, you can't mourn it.

The problem with this view is that there are plenty of people who are able to diagnose the epochal break, but celebrate, rather than mourn, it. When, for instance, the founders of the U.S. National Nanotechnology Initiative labeled nanotechnology "the next Industrial Revolution," they clearly had some kind of epochal break in mind, but only in the most positive sense. And when *The Economist* ("Innovation's Golden Goose" 2002) praised the Bayh-Dole Act as the "most inspired piece of legislation to be enacted in America" in the second half of the twentieth century (neglecting, as Steve Shapin [2003] has pointed out, the Civil Rights Act of 1964 and the Voting Rights Act of 1965, among others), they clearly meant it had opened a new epoch. After all, Bayh-Dole supposedly ushered in "a flowering of innovation unlike anything seen before." As David Edgerton (2007, ix) puts it in his critique of this way of talking: "We are told that change is taking place at an ever-accelerating pace, and that the new is increasingly powerful. The world, the gurus insist, is entering a new historical epoch as a result of technology. In the new economy, in new times, in our post-industrial and post-modern condition, knowledge of the present and past is supposedly ever less relevant."

The gurus' gushing doesn't indicate to me that diagnosis of the epochal break is sufficient to oppose it. In fact, it looks a lot like diagnosis is merely necessary to tout it. *Announcements* of epochal breaks have real consequences— they channel funding, they persuade people to join collaborations, they give bureaucracies a (wanted or unwanted) stimulus to reorganize. Such announcements puff up economic bubbles—remember the "new economy" of the 1990s, or Thomas Friedman's ubiquitously cited announcement that *The World Is* (now, all of a sudden, epoch-makingly) *Flat?* Climbing the hill to the point where we can view the consequences of announcing an epochal break is important. But continuing to climb to the point where the epochs look rather flat and continuous doesn't preclude skepticism about whatever epoch we are moving into. We need look no further than arch-modernist Bertolt Brecht to see that (as quoted by Edgerton 2007, viii):

> I stood on a hill and I saw the Old approaching, but it came as the New.
>
> It hobbled up on new crutches which no one had ever seen before and stank of new smells of decay which no one had ever smelt before.

I'm willing to admire the view from the epochal break vantage point. But I've not been convinced to linger there. Indeed, once I've reached the point that I can see an epochal break, I'm all the more eager to keep moving. One option is to head back down the mountain, to the kinds of microstudies that historians and sociologists of science thrive on. If the epochal break has any features worth studying, they should be visible, in some way, down at the microlevel of practice. The other option is to keep heading up the mountain—to view the current epoch in the context of its peers and thereby to historicize, and perhaps deflate, some of its ambitions and justifications.

The epochal break has some analytical value, in that historical actors often speak of their own eras as utterly different from all others. We should take those actors seriously—they have knowledge that we can never fully gain. Yet their claims of epochal specialness are interested claims—they are made for a purpose, to gather resources or convert foes into friends. As analysts, our interests differ from those of historical actors, as does the knowledge space that we have access to. We blind ourselves if we accept the claims for an epochal break without seeing the interests to which those claims are connected. We lame ourselves if we treat the putative present epochal break as so special or disruptive that we cannot learn from its resemblance to all of the other epoch-making turns, and all of the technoscience, that has come before.

REFERENCES

Biagioli, Mario. 1993. *Galileo Courtier: The Practice of Science in the Culture of Absolutism*. Chicago: University of Chicago Press.

Bridgman, Percy W. 1927. *The Logic of Modern Physics*. New York: Macmillan.

Edgerton, David. 2007. *The Shock of the Old: Technology and Global History since 1900*. New York: Oxford University Press.

Forman, Paul. 2007. "The Primacy of Science in Modernity, of Technology in Postmodernity, and of Ideology in the History of Technology." *History and Technology* 23: 1–152.

Galison, Peter, and Barton Bernstein. 1989. "In Any Light: Scientists and the Decision to Build the Hydrogen Bomb." *Historical Studies in the Physical and Biological Sciences* 19: 267–347.

Ghamari-Tabrizi, Sharon. 2000. "Simulating the Unthinkable: Gaming Future War in the 1950s and 1960s." *Social Studies of Science* 30: 163–223.

"Innovation's Golden Goose." 2002. *Economist*, December 14.

Kaiser, David. 2002. "Cold War Requisitions, Scientific Manpower, and the Production of American Physicists after World War II." *Studies in the Physical and Biological Sciences* 33: 131–59.

———. 2004. "The Postwar Suburbanization of American Physics." *American Quarterly* 56: 851–88.

Kirsch, Scott. 2005. *Proving Grounds: Project Plowshare and the Unrealized Dream of Nuclear Earthmoving*. New Brunswick, N.J.: Rutgers University Press.

Leslie, Stuart W. 1987. "Playing the Education Game to Win: The Military and Interdisciplinary Research at Stanford." *Historical Studies in the Physical Sciences* 18: 55–88.

Mirowski, Philip, and Esther-Mirjam Sent. 2007. "The Commercialization of Science and the Response of STS." In *The Handbook of Science and Technology Studies*, edited by Edward J. Hackett, Olga Amsterdamska, Michael Lynch, and Judy Wacjman, 635–89. Cambridge: MIT Press.

Shapin, Steven. 2003. "Ivory Trade." *London Review of Books*, September 11.

Wise, George. 1983. "Ionists in Industry: Physical Chemistry at General Electric, 1900–1915." *Isis* 74: 7–21.

Wise, Norton. 1988. "Mediating Machines." *Science in Context* 2: 77–113.

Breaking Up with the Epochal Break
The Case of Engineering Sciences

MIEKE BOON and
TARJA KNUUTTILA

EPOCHAL BREAKS ABOUND. After a short session on Google one learns that apart from the break between modernity and postmodernity, epochal breaks have taken place also between the eighteenth century and Romantic literature (Perry 1996), the oral Greek tradition and Hellenistic epic (Barnes 2003), African Christianity and its European roots, Ayatollah Ruhollah Khomeini's doctrine of *velayat-e faqih* and the traditional quietism of the Shi'i Muslims as well as between the different monetary standards (Cesarano 1999) and different sensorial topologies.[1] What this apparent abundance of epochal breaks tells us is that once a handy diagnostic term gets coined, it will be used to establish and discover patterns in the past, present, and even future, as Hans Radder suggests in this edited volume.

How then to evaluate the epochal break thesis concerning science or knowledge production? One obvious question concerns its timing. Most often the emergence of the brave new era of technoscience is located around the middle of twentieth century, becoming truly pervasive, however, first since the 1980s—a view with which Alfred Nordmann (in this collection) seems to agree. Gregor Schiemann, in turn—and to our mind rather plausibly—argues in his chapter that the transformation of science started already in the nineteenth century along with the development of "reality-shaping technologies."[2]

Instead of taking the phenomenon of epochal break at face value and then arguing whether or not it took place and if so, when, we take another route. We do not wish to engage in any large-scale historical argument but rather to critically examine the distinctions between representation and intervening, and basic and applied science that seem to be crucial to the "notion" of epochal break—at least as it is suggested by mode-2 literature. According to it, the epochal break between the two modes of research is seen as a breakaway from the theoretical, understanding-providing basic science toward interventional science taking place in the "context of application." As such, the epochal break, at least when it comes to scientific knowledge production, seems to reify those distinctions that have been criticized by the recent practice-oriented studies both in science and technology studies and in the philosophy of science.

This chapter looks at whether the aforementioned distinctions are tenable from the perspective of modeling in engineering sciences, which should provide the prime example of emerging technoscience. We argue that scientific modeling cuts interestingly across the two distinctions—that is, "representing versus intervening" and "basic versus applied"—thus challenging the basis of the epochal break thesis. Having found the basic distinctions on which the epochal break thesis rests untenable, at least when it comes to scientific practice, we ask what is in fact established by the epochal break talk.

Breaking Up the Divide between Representing and Intervening

In *Representing and Intervening,* Ian Hacking (1983, 145) claims: "Realism and anti-realism scurry about, trying to latch on to something in the nature of representation that will vanquish the other. There is nothing there. That is why I turn from representing to intervening." Statements like this have been taken to intimate that there is a fundamental divide between representation and intervention, the former intending to depict the world truthfully and the latter to manipulate things and shape the world. We propose, however, a different reading of Hacking. Namely, his claim that "there is nothing there" in the nature of representation hints to a more subtle stand concerning representation. Maybe there is nothing there *in our established notion of representation* that can accomplish what is expected from it. Thus taking a cue from Hacking, we suggest that the representational idiom does not describe adequately our actual theoretical practices, which instead of trying to accurately represent the world, are more tuned to our active engagement with the world than what is customarily supposed.

This entanglement of representing and intervening is particularly signifi-

cant for understanding scientific research in the context of technological applications. As we show later in the chapter, a distinction can be made between
engineering sciences, which primarily aim at scientific modeling, and *engineering*, which is more directly concerned with creating, producing, improving,
controlling, or designing various devices and materials. Yet no technoscientific
fusion of representation and technical intervention needs to follow from rejecting the customary representational idiom. But before going into this, let us
first visit the recent discussion on representation and modeling that testifies
to a certain intellectual insolvency concerning the traditional representational
paradigm *even* when it comes to the so-called basic science.

If we are to believe the philosophers of science, the link between models
and representation is as intimate as coming close to a conceptual one. Philosophers have generally agreed that models are essentially representations and
as such "models of" some real target systems. Yet the accounts given of the
representational character of models differ widely. The recent pragmatist approach to representation (e.g., Bailer-Jones 2003; Suárez 2004, 2010; and Giere
2004, 2010) could be seen as a critique of the structuralist notion that is part
and parcel of the semantic conception, which up until recently has been the
most widely held view on models. The semantic conception provides a straightforward answer to the question of how models give us knowledge of the world:
they specify structures that are posited as possible representations of either the
observable phenomena or, even more ambitiously, the underlying structures
of the real target systems. Thus, according to the semantic view, the structure
specified by a model represents its target system if it is either structurally iso-
morphic or somehow similar to it (e.g., van Fraassen 1980; French and Ladyman 1999; and Giere 1988). The pragmatist critics of the semantic conception
have argued, rather conclusively, that the structuralist notion of representation does not satisfy the formal and other criteria we might want to affirm of
representation (see, e.g., Suárez 2003; and Frigg 2002). The problem can be
located in the attempt to find such properties both from the representational
vehicle (the model) and the real object (the target system) by virtue of which a
representational relationship can be established between a model and its target
object.

So far, despite the numerous trials, no such even nearly adequate solution
to the aforementioned general philosophical puzzle concerning representation
has been presented. Hence the continued referral to representation as a putative
explanatory concept seems to be philosophically unfounded. The pragmatist alternative is to circumvent the traditional problem by making representational
relationship an accomplishment of representation-users. Consequently, what is

common among pragmatic approaches is the focus on the intentional activity of representation users and the denial that representation may be based only on the respective properties of the representative vehicle and its target object. However, if representation is primarily grounded in the specific goals and the representing activity of humans as opposed to the properties of the representative vehicle and its target, nothing very substantial can be said about it in general (see Giere 2004; and Suárez 2004).

Thus one largely ignored consequence of the pragmatic accounts of representation is the way they have emptied the concept of representation from much of its explanatory value. As a result, we are also deprived of an explanation of the epistemic value of models, too—that is, if we want to attribute it to representation. Consequently, Hacking 1983 can be read as not suggesting a turn away from representation but rather realizing that representation cannot accomplish what it is supposed to establish. If this were the case, how should one, then, approach the epistemic value of models? One obvious way out of this problem is not to attribute the knowledge-bearing properties of models to representation *alone*.

Interestingly, largely apart from the very interest in the topic of representation, a new discussion on models has emerged that loosens the epistemic value of models (at least partly) from representation and considers them as independent objects. The idea of models as independent objects or entities has been expressed by several recent authors in various ways. Margaret Morrison (1999) and Margaret Morrison and Mary Morgan (1999) have considered models as *autonomous agents* that are through their construction partially independent from theory and data. Michael Weisberg (2007) and Peter Godfrey-Smith (2006) in turn argue that models should be considered as independent entities in the sense of not representing any definite real target systems. According to them, modeling can be viewed as a specific theoretical practice of its own that can be characterized through the procedures of *indirect representation and analysis*. With indirect representation they refer to the way modelers construct simple, ideal model systems to which only a few properties are attributed, instead of striving to represent some real target systems directly.

How, then, are models as independent objects able to give us knowledge? Morrison and Morgan as well as Godfrey-Smith and Weisberg refer back to the notion of representation. But invoking representation would once again bring in the riddle of representation. In contrast, what we find the most important point in viewing models as independent things is that it enables us to appreciate their functional characteristics. Considering models from the functional perspective requires one to address them as *concrete objects* that are constructed in view of certain *epistemic purposes* and whose cognitive value derives largely from our *interaction* with them (Knuuttila and Merz 2009). Consequently, models can be considered as multifunctional *epistemic tools* (Knuuttila 2005; Knuuttila and Voutilainen 2003; Boon and Knuuttila 2009; Knuuttila and Boon forthcoming). From this perspective also the material embodiment of a model is of epistemic importance: the concrete representational means through which a model is achieved gives it the spatial and temporal cohesion that enables its manipulability. This also applies to so-called abstract models: when working with them we typically construct and manipulate external representational means such as diagrams or equations. Herein lays also the rationale for comparing models to experiments: in devising models, we construct self-contained artificial systems through which we can make our theoretical conjectures conceivable, articulated, and workable.

Indeed, it is rather paradoxical that the representational view on models actually abstracts away from the actual representational tools with which models are constructed. Assuming that theoretical activity seeks to depict the world *as it is,* it presupposes that we already knew what to represent and how—having also the appropriate means at hand for doing that. But the very point of representing (be it a model, piece of text, or some visualization produced by an inscription device) is to find out more about the phenomenon of interest, and it is more often than not guided by the available representational means (linguistic, pictorial, mathematical, and diagrammatical), computational templates and modeling methods at hand, not to mention the new media, such as computers. Models are artifacts, tools for articulating, finding out and bringing about rather than depicting the world truthfully.

It is important to note that the representational means used by sciences have different characteristic limitations and affordances; one can express different kinds of content with symbols than with pictures, for example. It is already a cognitive achievement to be able to express any hypothetical mechanism, structure, or phenomenon of interest in terms of some representational means. Such articulation enables further theoretical findings as well as new experimental setups, but it also imposes its own limitations on what can be done with a certain model. Yet the constraints built into a model also enable. As the real world is just too complex to study as such, models simplify or modify the problems scientists deal with. Thus modelers typically proceed by turning the constraints (e.g., the specific model assumptions) built into the model into affordances; one devises the model in such a way that one can gain understanding and draw inferences from using or "manipulating" it. This feature of models is linked to another characteristic of them. Especially theoretical models do importantly depict the possible and not just the actual. They are rather driven by pending scientific problems and questions than by the attempt to depict some real target system as accurately as possible.

We thus suggest that we typically learn from models by interacting with them—that is, by building them and trying out different things with them—which in turn explains why models are regularly valued for their *performance* and their *results* or *output*. From the functional perspective, rather than trying to represent some selected aspects of a given target system, modelers often proceed in a roundabout way, seeking to build hypothetical model systems in the light of their anticipated results or of certain general features of phenomena they are supposed to *bring about*. If a model gives us the expected results or replicates some features of the phenomenon, it provides an interesting starting point for further theoretical and experimental conjectures. The purposes the model is constructed for and the computability considerations often override in modeling the goal of realistic representation. Consequently, the very peculiarity of scientific modeling lies in its strategy of accounting for kinds of phenomena through the detour of constructing artificial entities in view of some scientific questions, keeping simultaneously in mind such pragmatic constraints as their tractability (see Humphreys 2004).

It seems to us that this conception of modeling fits especially well the engineering sciences because of their practical problem-orientation. The models in the engineering sciences are *not* first and foremost considered as accurate representations of some target systems, but rather as epistemic tools to find out how to produce, control, intervene—or to prevent some properties of materials or behavior of processes and devices. Consequently, engineering *scientists* build models for the purposes of imagining and reasoning about how to improve the performance of the devices, processes, or materials of interest. These models involve imaginable properties and processes, and they incorporate measurable physical variables and parameters (e.g., in the case of chemical engineering chemical concentrations, flow rates, temperature, and properties of materials such as diffusion, viscosity, and density). Often these models also incorporate dimensions of typical configurations of certain devices (see Boon and Knuuttila 2009).

Yet studying the modeling practice in engineering sciences should make it clear that although engineering sciences can be distinguished from engineering—thus not supporting any wholesale claim for technoscience—conceiving engineering sciences as basic research that provides theories that are then applied in concrete engineering tasks gives too simplified a picture of what happens in both domains. This simplified picture is of course what is to be expected if models are considered as truthful theoretical representations of the world. Then it is possible to assume that the success of their possible applications is due to their ability to depict the world truthfully (see Carrier's chapter in this edited volume). We believe instead that models produced in engineering science are not simply applied in engineering but used as epistemic tools. If we were to believe that models have the ability to be applied because they represent a target system accurately, it is often unclear which target system they seek to depict, especially in the context of engineering sciences. The devices, processes, or materials do not yet exist, but must be created—that is, they are designed by using models as epistemic tools. Models, on this account, allow for reasoning about (possible) interventions with the world—for instance, about how to generate a material property, how to make an efficient process, or how to avoid an undesirable phenomenon in a device.

Breaking Up the Divide between Basic and Applied

The way the mode-2 argumentation appears to rely on the already dated contrast between theoretical representation and intervening in the world makes us doubt the epochal break thesis concerning science already on philosophical grounds. What is more, we question also empirically whether any epochal break has taken place in how scientific research is actually done (see Gibbons et al. 1994; and Nowotny, Scott, and Gibbons 2003). We do not deny that there have been major institutional changes, or that there are important sociological aspects to the process of knowledge production. Neither do we doubt that political ideas on the role of science have changed in the past few decades. What we are critical about is the suggestion of mode-2 theorists that there has been a fundamental change within scientific practices regarding their goals, scientific methodologies, and epistemic criteria—that is, to how scientists think and write, and to the academic standards of good scientific research. In contrast, it seems to us that the way in which scientific theory is taught and used, the way in which experiments are designed, and the structure and content of scientific papers have been amazingly stable.

The break between knowledge production by theory (mode-1) and by application (mode-2) is largely justified by another distinction, which is rather inconspicuously intertwined with the divide between representing and intervening, as we have already hinted at. Namely, mode-2 literature presupposes that a clear distinction can be made between *basic* scientific research, which produces theories that meet academic standards, and *applied* scientific research, which produces knowledge that is highly context-dependent and that does not meet the classical epistemic criteria but instead solves problems. The epochal break thesis is founded on this divide. In a similar vein Martin Carrier and Alfred Nordmann (2006) claim that in present scientific research a markedly technological orientation can be observed that has had a significant impact on the goals and methodology of science. Theoretical representation is shifted to the background as scientific research focuses on useful properties and options for intervention. Rather than understanding nature, scientific research aims at shaping it. Again, it seems to us that this apparent change of orientation exhibits a change in how scientists justify their research but not in how scientific research is actually carried out. Indeed, this can be noticed in current research proposals, which are often written as if they were merely oriented at technological applications, making invisible the amount of scientific research that they actually entail.

If anywhere, the alleged recent emergence of the technoscience should most visibly be seen in the engineering sciences. The term "engineering science" goes back at least to the mid-eighteenth century in French, and the term in English was probably first used by W. J. M. Rankine, a Scottish engineer and scientist, in the 1850s.[3] Also later, in the late nineteenth and beginning of the twentieth century, researchers with an interest in engineering took scientific approaches to technologically produced phenomena (e.g., scientific researchers such as Sadi Carnot and Ludwig Prandtl). Their aim was scientific interpretation of phenomena occurring in, or produced by, technological devices, such as the production of power in heat engines examined by Carnot and flow phenomena in technological objects examined by Prandtl (see Boon 2006; and Boon and Knuuttila 2009). Engineering education has incorporated such scientific approaches, which has resulted in the present "engineering sciences." Today, engineering science manifests itself in hundreds of thousands of scientific researchers, thousands of scientific research laboratories, and hundreds of specialized scientific journals. Thus the engineering sciences more or less in their present form have existed already longer, yet the putative epochal break was observed first in the 1980s by most proponents of the thesis (see, however, Schiemann's chapter in this edited volume).

We suggest that the supposed break in scientific goals, methodology, and epistemic standards is due to what are taken as exemplary examples of research rather than to actual changes in how science is done. Philosophy of science, for instance, used to take physics as its exemplar—in particular its mathematical approach, which however was only a minor part of science. In stressing "the context of application," the mode-2 theorists in turn tend to take the application-oriented R&D departments in industry as their exemplar. This change of an exemplar of what counts as research does not legitimize the conclusion that a fundamental change in scientific goals, methodology, and criteria has taken place.[4]

As such, the distinction between basic science and applied science is much older than mode-2 theory of knowledge production. Many researchers themselves and especially the popular accounts of science tend to take for granted a distinction between basic scientific research that aims at theories and understanding nature, and applied science that aims at shaping nature by intervention. Wikipedia, for instance, presents the following description of applied physics:

> *Applied physics* is a general term for physics which is intended for a particular technological or practical use. "Applied" is distinguished from "pure" by a subtle combination of factors such as the motivation and attitude of researchers and the nature of the relationship to the technology or science that may be affected by the work. It usually differs from *engineering* in that an applied physicist may not be designing something in particular, but rather is *using physics or conducting physics research with the aim of developing new technologies or solving an engineering problem.* . . . In other words, applied physics is rooted in the fundamental truths and basic concepts of the physical sciences but is concerned with the utilization of these scientific principles in practical devices and systems.[5]

This definition reflects how branches of the engineering sciences are commonly understood: the divide between "applied" and "pure" leaves engineering sciences on the side of applied science. It suggests that scientific principles are simply used in the development of better devices. Although scientists also frequently present their work in this way, they might not give a good characterization of it, as scientists often lack appropriate meta-theoretical means for observing and describing their research, causing the traditional picture to hold them captive. We suggest an alternative to this picture: the conception of models as epistemic tools that fits engineering sciences as well as the "pure" sciences. This is because the possible differences between the two lie rather in the phenomena they are interested in than in their epistemic practices. The engineering sciences develop models of phenomena relevant to technological applications, which, however, are not less "deep" than models of phenomena produced in "pure" sciences.

Scientific articles provide a good indicator of the prevalent scientific goals, methodology, and epistemic criteria of engineering sciences. When browsing through journals in different fields of the engineering sciences, one quickly learns that scientific models are at the center of interest. These models aim at yielding scientific understanding of the behavior of devices or the properties of materials. A classic example is how Sadi Carnot in the early nineteenth century translated the functioning of heat engines into a theoretical problem concerning the phenomenon of producing motion by heat. This led to conceptual novelties and contributed to the consequent development of the thermodynamic theory. Thus scientific research pertaining to the functioning of devices need not be fundamentally different from what is commonly understood as "pure" science (see Boon and Knuuttila 2009; and Knuuttila and Boon forthcoming).

Scientific research in chemical engineering provides another example. Very similar to Carnot's approach, scientific researchers in this field proceed by studying the behavior of devices by interpreting them in terms of physical phenomena considered to be relevant to their proper or improper functioning, and then modeling these phenomena. Examples of such modeled phenomena are desirable and undesirable chemical reactions; the transport of liquids, gases, and solids within the device; the transport of chemical compounds by means of fluid flow or diffusion in the fluid; the transport of heat by convection or conduction; and other physical processes, such as absorption, dissolution, ionization, precipitation, vaporization, and crystallization. In modeling these physical phenomena, usually two different types of scientific models are produced simultaneously: models that present a causal-mechanistic understanding of the phenomenon, and models that present a mathematical description of the phenomenon in terms of relevant physical variables—in this they also resemble other natural sciences (Boon 2006, 2009).

Consequently, it seems to us that the engineering sciences do not fit very well either with mode-2 or with mode-1. In particular, the fact that the engineering sciences aim at *understanding* phenomena makes them fit better with mode-1. On the other hand, these phenomena are related to the functioning of devices or materials, which in turn better suits mode-2. Are engineering sciences then perhaps basic or applied, or neither? We propose that engineering sciences make a case for rejecting the divide between basic and applied sciences. Their scientific goal is usually related to concrete problems, which is in accordance with mode-2, but within scientific practice their goal is narrowed down to modeling a phenomenon, which aims at understanding it scientifically. That they should proceed in this way blurs any clear distinction between them and the putative basic sciences, which also aim at getting theoretical grasp of phenomena by modeling.

Conclusion

We have argued that the very distinctions on which the alleged epochal break between mode-1 and mode-2 is built do not hold up to closer philosophical or empirical scrutiny. Moreover, neither do they conform even to engineering sciences that should provide the closest case of what is commonly conceptualized as technoscience. This should come as no surprise since most of the claims concerning an epochal break are not based on empirical studies. As any representations are purpose-relative, one might ask what are the purposes for representing the current scientific activity as if an epochal break has taken place there. For us it seems that this claim should be evaluated rather as a piece of political rhetoric than any serious claim about scientific practice. Or, to give it a more favorable reading, one could approach epochal break and the associated modes as *transdiscursive concepts* that are used to reorganize and guide discourses between the research communities and policy making.

Transdiscursive concepts aim to mobilize and empower as well as to create social consensus (see Miettinen 2002, 132–41). From this perspective the critical question concerning the alleged epochal break pertains to whether it really succeeds in its self-renowned task of empowering the problem-solving activity regarding the complex problems of the twenty-first century. We doubt that this is the case, if only because the vision of scientific practice embedded in it is naïve to say the very least—as we have argued in this chapter. Thus we doubt whether any remarkable problem-solving ability is going to result from all those political science-policy measures that are taken in the name of transdisciplinarity. The mode-2 talk grossly overestimates what scientists can do in the short time spans—that is, to which extent innovative scientific research can be geared to practical problem-solving—and last but not least, it underestimates the real communicative and other difficulties involved in interdisciplinary work. This is not to say that ideologies like these would not have an effect on academic life. They do and we already see them around.

Mode-2 and related policies have offered governmental and business sectors as well as administration and various funders a good justification to exert more pressure on the academic sector in terms of (short-term) accountability and commodification. Partly as a result of these pressures, universities have turned into "multiversities" (Kerr 1966) that have taken multiple tasks upon themselves, suffering simultaneously from the dearth of funding. As universities have sought to renew their financial base through contract research, educational services, consulting, and the commercialization of research results the mode-2 ideology legitimizes the status quo by offering a rosy vision of the organizational and other changes that are taking place. We wonder if this development is rather worsening the conditions for any truly innovative research.

NOTES

1. On African Christianity and its European roots, see Bulletsandhoney, "Signs the Devil Roams among Us and That the Kenyon Nation Shall Be Born in Church," *African Bullets & Honey* (blog), December 13, 2005, online at http://bulletsand honey.wordpress.com/2005/12/. Also see the editorial in *Middle East Report* ("The Shi'a in the Arab World" theme issue) 37 (242), from the Middle East Research and Information Project, online at http://www.merip.org/mer/mer242/editorial.html.
2. However, this way of doing science emerged already when scientists started to use instruments in their experiments—that is, earlier than what Schiemann claims (see also Radder's chapter in this edited volume).
3 We wish to acknowledge David Channell, who gave us valuable information on the history of the notion of "engineering sciences."
4. It is noteworthy that R&D has not been regarded as a scientific research practice in the past accounts of science. Yet even R&D departments often do scientific research, including the publishing of scientific articles. However, as already mentioned, an important difference lies in how they present their purposes, which is partly determined by the audience to whom they have to justify their activities, such as their company, their peers, the government, the general audience, and so on.
5. Wikipedia, "Applied Physics," online at http://en.wikipedia.org/wikiApplied _physics.

REFERENCES

Bailer-Jones, Daniela. 2003. "When Scientific Models Represent." *International Studies in the Philosophy of Science* 17: 59–74.

Barnes, Michael. 2003. "Oral Tradition and Hellenistic Epic: New Directions in Apollonius of Rhodes." *Oral Traditions* 18 (1): 55–58. Online at http://journal. oraltradition.org/files/articles/18i/Barnes.pdf.

Boon, Mieke. 2006. "How Science Is Applied." *International Studies in the Philosophy of Science* 20 (1): 27–47.

———. 2009. "Understanding in the Engineering Sciences: Interpretative Structures." In *Scientific Understanding: Philosophical Perspectives*, edited by Henk de Regt, Sabina Leonelli, and Kay Eigner, 249–70. Pittsburgh: University of Pittsburgh Press.

Boon, Mieke, and Tarja Knuuttila. 2009. "Models as Epistemic Tools in Engineering Sciences: A Pragmatic Approach." In *Philosophy of Technology and Engineering Sciences*, edited by Anthonie Meijers, 687–719. Amsterdam: Elsevier Science.

Carrier, Martin, and Alfred Nordmann. 2006. "Science in the Context of Application: Methodological Change, Conceptual Transformation, Cultural Reorientation."

Application for a ZiF-Research Group at the Center for Interdisciplinary Research (Zentrum fur Interdisciplinäre Forschung, ZiF). Bielefeld, Germany.

Cesarano, Filippo. 1999. "Competitive Money Supply: The International Money System in Perspective." *Journal of Economic Studies* 26 (3): 188–200. Online at http://www.emeraldinsight.com/journals.htm?articleid=846108&show=html.

French, Steven, and James Ladyman. 1999. "Reinflating the Semantic Approach." *International Studies in the Philosophy of Science* 13 (2): 103–21.

Frigg, Roman. 2002. *Models and Representation: Why Structures Are Not Enough.* Measurement in Physics and Economics Discussion Paper Series. London: London School of Economics.

Gibbons, Michael, Camille Limoges, Helga Nowotny, Simon Schwartzman, Peter Scott, and Martin Trow. 1994. *The New Production of Knowledge: The Dynamics of Science and Research in Contemporary Societies.* London: Sage.

Giere, Ronald N. 1988. *Explaining Science: A Cognitive Approach.* Chicago: University of Chicago Press.

———. 2004. "How Models Are Used to Represent Reality." *Philosophy of Science* (symposia) 71: 742–52.

———. 2010. "An Agent-Based Conception of Models and Scientific Representation." *Synthese* 172: 269–81.

Godfrey-Smith, Peter. 2006. "The Strategy of Model-Based Science." *Biology and Philosophy* 21: 725–40.

Hacking, Ian. 1983. *Representing and Intervening.* Cambridge: Cambridge University Press.

Humphreys, Paul. 2004. *Extending Ourselves: Computational Science, Empiricism, and Scientific Method.* Oxford: Oxford University Press.

Kerr, Clark. 1966. *The Uses of the University.* New York: Harper & Row Publishers.

Knuuttila, Tarja. 2005. "Models, Representation, and Mediation." *Philosophy of Science* 72: 1260–71.

Knuuttila, Tarja, and Martina Merz. 2009. "Understanding by Modeling: An Objectual Approach." In *Scientific Understanding: Philosophical Perspectives,* edited by Henk de Regt, Sabina Leonelli, and Kay Eigner, 146–68. Pittsburgh: University of Pittsburgh Press.

Knuuttila, Tarja, and Atro Voutilainen. 2003. "A Parser as an Epistemic Artefact: A Material View on Models." *Philosophy of Science* 70: 1484–95.

Knuuttila, Tarja, and Mieke Boon. Forthcoming. "How Do Models Give Us Knowledge? The Case of Carnot's Ideal Heat Engine." *European Journal for Philosophy of Science.*

Miettinen, Reijo. 2002. *National Innovation System: Scientific Concept or Political Rhetoric?* Helsinki: Edita.

Morrison, Margaret. 1999. "Models as Autonomous Agents." In *Models as Mediators: Perspectives on Natural and Social Science,* edited by Mary S. Morgan and Margaret Morrison, 38–65. Cambridge: Cambridge University Press.

Morrison, Margaret, and Mary S. Morgan. 1999. "Models as Mediating Instruments." In *Models as Mediators: Perspectives on Natural and Social Science,* edited by Mary S. Morgan and Margaret Morrison, 10–37. Cambridge: Cambridge University Press.

Nowotny, Helga, Peter Scott, and Michael Gibbons. 2003. "Introduction—'Mode 2'
   Revisited: The New Production of Knowledge." *Minerva* 41: 179–94.
Perry, Seamus. 1996. "Review of *Wordsworth's Pope: A Study in Literary Historiography*,
   by Robert J. Griffin." *Romanticism on the Net* 4. Online at http://www.erudit.org/
   revue/ron/1996/v/n4/005728ar.html.
Suárez, Mario. 2003. "Scientific Representation: Against Similarity and Isomor-
   phism." *International Studies in the Philosophy of Science* 17: 225–44.
———. 2004. "An Inferential Conception of Scientific Representation." *Philosophy of
   Science* (symposia) 71: 767–79.
———. 2010. "Scientific Representation." *Blackwell's Philosophy Compass* 5: 91–101.
van Fraassen, Bas. 1980. *The Scientific Image.* Oxford: Oxford University Press.
Weisberg, Michael. 2007. "Who Is a Modeler." *British Journal for the Philosophy of
   Science* 58: 207–33.

Science and Its Recent History
From an Epochal Break to Novel, Nonlocal Patterns

HANS RADDER

THE EPOCHAL BREAK THESIS COMES in several versions. What they have in common is the claim that during a limited period of time, science, as it is actually practiced, has changed substantially or even essentially. Moreover, this change is taken to mark the start of a new age. That is to say, its occurrence is intrinsically related to a wide-ranging, epoch-making sociocultural development and its impact extends far beyond the specialized practices of the sciences. Most advocates of the thesis agree about the dating of the break: the year 1980 is often mentioned as a focal point, even if this dating is as often (and rightly) qualified by pointing to developments in the earlier decades as leading up to the break. Beyond this general characterization, we find different articulations of the thesis. In his contribution to this edited volume, Alfred Nordmann sees the essential change in a transition from a scientific enterprise to a regime of technoscience. The primary focus of this view is on the distinct philosophical status of science as compared to technoscience. The mode-1/mode-2 approach, briefly summarized in Gregor Schiemann's contribution in this collection, has a much stronger focus on institutional, economic, and policy features. Mode-1 research is claimed to be autonomous, academic, disciplinary, and methodological, while mode-2 knowledge production is characterized by taking place in application contexts, by being commercialized and transdisciplinary, and by essentially including social criteria of accountability and quality control (see Gibbons et al. 1994).

The epochal break thesis constitutes a bold claim with historical, philosophical, social, and moral dimensions. This chapter discusses some aspects of each of these dimensions. First, I argue that the idea of a single "great divide" between a scientific enterprise and a regime of technoscience is questionable on both historical and philosophical grounds. Yet this does not imply that there are no important distinctions at all between recent and past science. In section 2, I point to two novel patterns that can be distinguished in recent scientific practices: a strong focus on the issue of the external validity of scientific methods and claims, and a significant commodification of academic research. In section 3, I conclude that a conception of scientific development in terms of the emergence and reproduction of novel, nonlocal patterns is preferable to an account in terms of an epochal break. Furthermore, using Max Weber's ideal-type approach, I provide a sketch of how nonlocal patterns may be identified and explained, and what is implied, as well as what isn't, in postulating the existence of such patterns. The chapter closes with an argument for making explicit the normative issues involved in advocating philosophical claims—be they about epochal breaks or about novel, nonlocal patterns. In the present case, this implies highlighting, scrutinizing, explaining, and assessing the implications of the focus on external validity and of the commodification of academic research.

1. No Great Divide between Science and Technoscience

Let me start with history. An important part of the arguments for or against an epochal break pertains to alleged or disputed changes in the relationship between science and technology. Obviously, these arguments presuppose some account of the relationship between science and technology before and after the break. Unfortunately, these accounts are often less than adequate. Thus in his chapter in this edited volume Schiemann takes for granted the appropriateness and correctness of the "classical conception of science," in which scientific knowledge is characterized in terms of generality, necessity, and truth. This classical conception, however, is primarily a creation of philosophers, which is very hard to reconcile with science "as it was actually practiced." The fact that this conception had a certain impact on the self-understanding of certain scientists (and hence *some* impact on their practice) by no means implies that it can be taken for granted as a plausible *general* account of science.

Thomas Kuhn (1977) has characterized the development of the physical

sciences between the seventeenth and the mid-nineteenth century in terms
of two, largely independent traditions: a mathematical or classical and an
experimental or Baconian tradition. The mathematical tradition was a
transformation and extension of ancient sciences that included astronomy,
harmonics, mathematics, optics, statics, and the study of motion. The Baconian
tradition emerged in the seventeenth century and focused on the study of
pneumatic, magnetic, electrical, chemical, and heat phenomena. It was only
in the mid-nineteenth century that these two traditions became systematically
joined through the mathematization of the study of such phenomena as heat,
light, electricity, and magnetism. Even if we grant (but only for the sake of
argument) that the classical conception of science applies to the mathematical
tradition, it is obvious that it does not hold for the Baconian sciences.[1] These
sciences exhibited several features that are prima facie technoscientific in
nature: they addressed a diversity of disconnected empirical domains; they
depended essentially on the invention, use, and improvement of technological
instrumentation; and their experimental approach implied the explicit
significance of practical intervention.

At a smaller scale, these two traditions can be found in the debates between
Robert Boyle and Thomas Hobbes about the science and philosophy of Boyle's
air-pump experiments. In these debates, which took place in England in the
1660s and early 1670s, Hobbes represented the classical conception of science
with its emphasis on first principles, general truth, and certainty. Boyle, on
the other hand, stressed that knowledge of matters of fact was only probable,
never certain; his reasoning was empirical, not based on first principles; and
the phenomena realized by means of the air pump were clearly artificial:
the "empty" space within the air pump constituted an artificially created
phenomenon. As Steven Shapin and Simon Schaffer (1985, 26) state: "We
start by noting the obvious: matters of fact in Boyle's new pneumatics were
machine-made."

Against what is suggested by Shapin and Schaffer's interpretation, I do
not want to put forward Boyle and the advocates of the Baconian sciences as
the final victors in these debates. The mathematical sciences continued to
exist and flourish between the seventeenth and mid-nineteenth centuries.
Yet so did the Baconian sciences, and if we would move from the physical
sciences to the earth sciences, the biomedical sciences and the engineering
sciences, their relative significance would even increase. My conclusion is that
seeing theoretical representation of nature as typical of a scientific enterprise
and practical intervention as characteristic of a regime of technoscience is
inappropriate because of the actual role of the interventionist practices of the

Baconian sciences (see also the account of the engineering sciences by Mieke Boon and Tarja Knuuttila in this edited volume).

In a way this is conceded by some of the proponents of the epochal break thesis. Nordmann, for example, admits that technology, in the form of instrumentation, plays an important role in the scientific enterprise. He claims, however, that the point of his version of the epochal break thesis is more subtle. In the scientific enterprise it is still possible to separate conceptually the scientific (that is to say, the natural) objects from the technological (that is, the artificial) objects. In contrast, in the regime of technoscience this "purification work" is claimed to be "no longer possible and no longer required."

I think, however, that this subtlety cannot rescue the thesis. First, it is important to see that the claim that there is, or is not, a conceptual separation between the natural and the artificial is itself a result of philosophical interpretation. This means that the subtle thesis does not bear any more upon science "as it is actually practiced" but depends on the plausibility of a specific philosophical interpretation of this practice and its products.[2] Second, if this is the case, there is the problem of the existence of alternative philosophical interpretations of the scientific enterprise and the regime of technoscience. For instance, at a general level Wittgensteinian philosophers have long argued that the classical epistemological "search for certainty" has always been a failure, completely independent of whatever changes may have, or may not have, occurred in the history of science (Hamlyn 1970, 10–16).

More specifically, the accounts by the social constructivists Shapin and Schaffer and by the general constructivist Bruno Latour imply that the "scientific enterprise" has always been technoscientific. Thus, on these alternative interpretations, the idea of an epochal break makes no sense. Conversely, one can argue, as I did in Radder 1996 (chapter 4), that a distinction between human-independent, natural potentialities and human-dependent, historical realizations, and more specifically between reproducible and reproduced experiments and observations, can be applied to both the scientific enterprise and the regime of technoscience. Hence, on this interpretation purification is still possible. Finally, if a distinction between subjective and objective aspects of the generation and content of scientific knowledge is no longer required, how should we explain the occurrence, the fierceness, and the impact of the recent "science wars" that started in the early 1990s and lasted until the early years of the new millennium (see Brown 2001)? After all, one prominent feature of these extensive and heated controversies was the sharp defense of the objectivity of science by both scientists and philosophers of science. In sum, the general conclusion of this section is that there are good reasons for questioning the occurrence of a great divide (both in a plain and in a more subtle sense) between science and technoscience.

2. Two Novel Patterns: External Validity and Commodified Research

Yet, what I have said so far does not mean that current science is business as usual. Important changes have recently taken place. Let me mention two important novel patterns.[3] The first is the *significance of the external validity* of scientific knowledge claims and methods. Scientific knowledge and methods are often developed within artificially restricted contexts, such as laboratory experiments, field tests, modeling practices, closed surveys, and the like. Internal validity is the validity of this knowledge and these methods for those contexts. For instance, a certain claim in economics may be internally valid for a theoretical model of a specific type of transaction among idealized economic agents; or a biomedical claim may correctly apply to the in vitro effectiveness of a certain vaccine in a laboratory experiment.

Present-day science, however, is increasingly interested in the validity of its methods and claims beyond such internal contexts. Policy makers want to know the economic behavior of real people, and doctors need pharmaceuticals that are effective in vivo, for real patients. In many cases, the external or "real-world" situations are much more *complex* and much *harder to control* than the corresponding internal situations (see also Radder 1996, chap. 6, and the chapter by Astrid Schwarz and Wolfgang Krohn in this edited volume). For this reason, the extrapolation of internally valid results to complex and hard-to-control external situations often proves to be difficult. The theoretical models turn out to be poor predictors of real-world economic behavior, and the laboratory vaccines do not work when administered to real human patients. Hence the novel pattern involves the attempts to investigate and ascertain the external validity of the scientific methods and knowledge claims. Typically, this "coming out" of science requires the inclusion of a broad range of relevant factors, a more interdisciplinary approach, the involvement of both scientific and nonscientific actors, and a keen eye for uncertainties, limitations, and social and moral problems of the uses of scientific knowledge and methods. An exemplary illustration is the research on climate change and its natural and social implications coordinated by the Intergovernmental Panel on Climate Change (see Petersen 2006).

Philosophers of the social and human sciences have acknowledged the distinction between internal and external validity for some time. Yet the nagging question has always been, and it still is, what internal results obtained under specific conditions and hence applying to a limited domain tell us about human beings in their daily lives or about social structures in the outside world. Thus far, in the philosophy of the natural sciences the issue has been acknowledged far less widely. It is to the credit of Roy Bhaskar and later Latour to have recognized the importance of the problem of external validity for the natural sciences. Bhaskar (1978 [1975]) coined the contrast between closed and open systems, and his account of the natural sciences implies that their external validity in open systems is limited to post-hoc explanation. Latour (1983) admitted that external validity is harder to come by than internal validity, but claimed that it can be achieved if one succeeds in recreating the laboratory conditions in the outside world.[4]

Nevertheless, in spite of Bhaskar and Latour, the claim that internally valid results are "in principle" also externally valid has been, and to a certain extent still is, a taken-for-granted axiom of many philosophers of the natural sciences and, in their wake, of the self-understanding of many natural scientists. In this respect the classical conception of science, in particular its claimed generality, has had a substantial impact on the understanding of science by philosophers and scientists. Furthermore, this conception was often complemented by a view that separates science from technology. Of course, these philosophers and scientists realized that achieving external validity in practice is often difficult, but this task—and this is the important point—was not perceived as a task for the scientists anyway but rather for the engineers and technologists (see Forman 2007). Hence, if a particular demonstration of external validity failed, it was not the axiom of natural science but technology or engineering that was to be blamed.

The increased focus on the issue of external validity is related to the "scientification of society" and the "socialization of science." Increasingly, technology, medicine, politics, war, education, jurisdiction, and so on make an appeal to scientific methods or scientific knowledge; scientists, in turn, actively anticipate and respond to such appeals. Hence, we are witnessing a dual process of the scientification of society and the socialization of science. Consider the recent winners of the Spinoza prize, the highest award for scientific excellence in the Netherlands, sometimes called the Dutch Nobel prize. The research topics of the four 2008 winners are telling enough: diagnostic methods for rare diseases in children, nanotechnological techniques for increasing the speed of computer memories, production of dairy products with improved taste and more vitamins, and the cultural history of European nationalism.

The connection with the processes of scientification and socialization adds an important qualification to the issue of external validity. Lack of external validity often results from substantial uncertainties in scientific claims and methods. Such uncertainties are said to be a central characteristic of postnormal science. Some traditional disciplines, however, also address complex and hard-to-control external systems, and hence they have to deal with large uncertainties. As a student of physics, I remember being struck by the fact that the theoretical and empirical uncertainties that were seen to be reasonable within the field of astrophysics were enormous as compared to what was tolerated in other fields of physics. Yet astrophysicists could live with these uncertainties, presumably because no direct social stakes were involved. Apparently, a big uncertainty about the life course of a particular type of star matters a lot less than the same uncertainty in a model of global climate change. Thus, even if the existence of large uncertainties is not limited to postnormal science, the strongly increased scientification of society and the corresponding socialization of science imply that the significance of these uncertainties is now seen to be much greater.

A second novel pattern is the *commodification of academic research* (see Radder 2010). By "academic research" I mean, primarily, research in public universities and other publicly funded research institutes (I add "primarily" because one could argue that academic research may also be carried out at nonprofit, yet private, universities). In relation to the epochal break thesis, speaking of *academic* science implies a qualification and limitation of this thesis. More strongly, it suggests that this thesis had better be limited to academic science right from the start. "Commodification" is a complex phenomenon. It denotes any process in which scientific activities and their results are predominantly interpreted and assessed on the basis of economic criteria.[5] In this sense, commodification includes not only contract research commissioned and paid by commercial organizations, but also the increasing corporatelike management and culture of the university, the political reduction of the "social relevance of science" to "its potential for economic innovation," and the like. The last point shows that, empirically, commodification is associated with the presently dominant form of scientification and socialization. Yet, conceptually and normatively, the option of a noncommodified scientification and socialization should be fully acknowledged.

How widespread and influential is the commodification of academic research? In the past years some studies of this question have been made, primarily by concerned journalists, policy makers, and scientists (see, e.g., Bok 2003; Krimsky 2003; and Slaughter and Rhoades 2004), but much more could and should be done. Thus far, philosophers of science and science studies scholars have hardly begun to explore and scrutinize this subject. Yet, both the available studies and my own experience as a longtime researcher in an academic institution support the claim that commodification is a wide-ranging phenomenon with a significant impact.

Quite a few studies focus on several highly publicized negative impacts of commercial influences on the quality of scientific knowledge, in particular in the biomedical sciences (see also James Robert Brown's chapter in this edited volume). This has been, and continues to be, important because it raises the general awareness of the issue of commodification. Yet, more pervasive are the structural transformations of academic life. Just consider the following examples:

1. The changing academic culture (for example, independent university periodicals as vehicles for critical debate on all kinds of academic matters become part of a centrally controlled PR strategy, devised by top university administrators and managers without an academic background).

2. The increasingly common entrepreneurial ethos (for example, researchers can be part-time employed by the university and part-time running their private business in the same area as their academic research).

3. Changing socioepistemic norms (for example, quite generally acquired patents—that is to say, commercial monopolies—come to be seen as fully legitimate academic achievements, comparable to journal publications).

How do these two patterns (the focus on external validity and the commodification of academic research) relate to Nordmann's account of a regime of technoscience? Surely, several aspects of dealing with complex and hard-to-control external systems can be fitted into the idea of a technoscience. One may, for instance, think of the application of computational and simulation methods (see Ann Johnson and Johannes Lenhard's chapter in this edited volume). Indeed, such methods often aim at mastering complexity and uncontrollability, and they constitute a prominent model of the intrinsic entanglement of the natural and the artificial that is claimed to be characteristic of technoscientific research. Other aspects of this pattern—in particular, the involvement of nonscientific actors and the economic and sociocultural issues—are not taken into account in the idea of a regime of technoscience. This point applies even more to the pattern of the commodification of academic research. Thus Nordmann's philosophical approach seems to have bracketed the economic and sociocultural dimensions of the changes in the historical development of the sciences. In this respect there is a clear contrast with the mode-1/mode-2 approach, which provides a more comprehensive description of these economic and sociocultural developments.

3. From an Epochal Break to Novel, Nonlocal Patterns

It is hard to deny that the strong focus on complex and hard-to-control external problems and the commodification of academic research constitute important patterns of present-day science. Somewhere between 1960 and 1980 these phenomena have really acquired a major significance in actual scientific practices, and they have maintained this significance up to the present day. Thus far, this is plausible enough, but is it enough to support an epochal break thesis? At this point, and in line with the arguments in section 1 of this chapter, it is important to acknowledge the historical and empirical criticism of the novelty claim—for instance, as represented in Cyrus Mody's chapter in this edited volume.[6] That is to say, academic commodification and focusing on external validity is not strictly new. It is also the case, however, that even if the phenomena as such can be found in earlier science, their incidence and significance have increased substantially in the period under discussion. This justifies the use of the notion of patterns and indeed of novel patterns. Hence, in line with Valerie Hanson's chapter in this edited volume, my conclusion is that the changes are a matter of intensification rather than complete transformation.

The fact that the historical and empirical criticism has a point requires that both the proponents of an epochal break and the advocates of the rise of novel patterns need to reflect on the methodological issue of how to substantiate their claims. I suggest to interpret the phenomena of commodification or of seeking external validity as constituting a pattern, and more specifically a "nonlocal" pattern. Such nonlocal patterns are not exceptionless regularities but rather substantial trends that require material and social work for being produced and maintained and that possess a broad significance that by far transcends the significance of each separate local realization of this trend (see Radder 1996, 101–5 and 169–74; and Radder 1997, 649–52). As illustrations, I referred to Kuhn's pattern of normal science, crisis, and revolution, to Alistair Crombie's and Ian Hacking's styles of reasoning, and to the strongly increased militarization of science during the twentieth century.

Just like observing a pattern in a certain painting or a piece of tapestry, observing a nonlocal pattern in scientific development requires being in the right position and having the right perspective. Conversely, it is easy not to see this pattern by looking more closely at the details of the historical episodes. Moreover, by watching from another distance or taking a different perspective, one may discern other patterns. Yet it is important to appreciate that this dependence on specific observational conditions does not make nonlocal patterns unreal or subjective: the pattern is there to see for anybody who is (or might be) in the right position and takes (or might take) the appropriate perspective. The advantage of seeing history as nonlocally patterned is that it avoids both the questionable postmodernist view of history as a rambling bag of fully disconnected, local events and Paul Forman's (2007, 69) exaggerated claim that "if . . . we reject the notion of historical eras as a modernist illusion, or, worse, oppression, we are inevitably also rejecting history as a scholarly discipline."

Having identified certain nonlocal patterns in the history of science or technology, we may proceed to explaining the rise and reproduction of these patterns. Here, the most balanced approach is still Max Weber's theory of explanation by means of ideal-types. Ideal-types are theoretical concepts that play a key role in explanatory hypotheses about the historical episodes in question. Ideal-types are not normative ideals but concepts that enable a clear and illuminating structuring and accentuation of historical events and episodes.[7] As suggested in the previous section, one may conjecture that "the focus on the issue of external validity" could be explained by hypotheses that include the ideal-typical concepts of the "scientification of society" and the corresponding "socialization of science"; similarly, an explanation of "the commodification of academic research" could exploit the ideal-typical concept of the "rise of neoliberal worldviews, politics and policies."

I think that these methodological remarks are not incompatible with Nordmann's position. If so, however, two important qualifications should be added. The first pertains to the appropriateness of the notion of a break. On the one hand, seeing the rise of a technoscientific regime (or, more broadly, a mode-2 science) as a collection of emerging, nonlocal patterns that need further empirical articulation and ideal-typical explanation is certainly preferable to seeing it as a straightforward empirical generalization, which seems to be the usual interpretation in historical and sociological accounts of this subject. On the other hand, we have seen that the reality of such patterns does not necessarily contradict the simultaneous existence of other nonlocal patterns, and here lies the partial right of Schiemann's alternative periodization. An implication of the possibility and actuality of the coexistence of distinct nonlocal patterns is that the metaphor of a "break" is inappropriate. Thus, even if one grants the rise of neoliberal worldviews and politics or the processes of scientification and socialization "epochal" significance (which I am inclined to do), it does not follow that academic commodification or a focus on external validity constitutes a single great divide between recent and earlier science.

A second, and a more consequential, qualification of the epochal break position derives from a further important aspect of Weber's theory of explanation.

According to this theory, the initial selection of particular ideal-types, rather than different ones, implies a value-laden stance by the investigators or their communities concerning the historical significance of the phenomena covered by these ideal-types. Here we touch the normative and reflexive dimensions of philosophical explanation and interpretation (Radder 1996, chap. 8; and Radder 1997, sec. 6). An explicit acknowledgment of the significance of these dimensions and, more important, a reasoned normative and reflexive assessment of the alleged epochal break is lacking from Nordmann's chapter. The same applies, albeit to a smaller extent, to Schiemann's contribution.

In contrast to these authors (and in this respect going beyond Weber), I think that an important aspect of a comprehensive philosophical interpretation is to provide such a reasoned assessment. For example, there are good reasons for judging several implications of the commodification of academic research to be highly questionable. These reasons are both epistemic and social or moral. Elsewhere I have collected a variety of comprehensive and in-depth discussions of the intricate normative issues of academic commodification (see Radder 2010). In this brief chapter I can only mention a few of its problems, following the three examples provided in the preceding section:

1. A commodified academic culture leads to an erosion of the space for open, critical debate (see, for instance, the recent attempts by university administrators to transform independent university periodicals into docile PR magazines).

2. An entrepreneurial ethos entails increased conflicts of interest and risks of abuse of public resources and money (for example, in the case of combining a university appointment with running a private firm in the same area, a type of arrangement that would be utterly illegitimate for journalists, politicians, or judges).

3. Compared with the standard review processes of academic journals, the disclosure of the research results in patenting applications is far less detailed and the assessment of their epistemic value far less thorough.

From the perspective of this chapter, highlighting, scrutinizing, explaining, and assessing nonlocal patterns, such as the commodification of academic research, is of prime significance. After all, since the rise of nonlocal patterns is not based on necessary laws of history, they are capable of being reproduced or strengthened but also of being weakened or changed in favor of normatively more desirable future patterns.

NOTES

1. An apparent problem of Kuhn's account is that at least some of the disciplines in the classical tradition (for instance, optics and mechanics) did include substantial experimental research.
2. Of course, our descriptions of the practice of science include interpretation as well; yet this kind of interpretation can, and should, be distinguished from the more general philosophical interpretation of what these descriptions mean. Hence Forman (2007, 6 and 10) rightly distinguishes between the practice of science and technology and their cultural or philosophical interpretation, and he consistently restricts his main claims to the epochal change in the *interpretation* of the relationship between science and technology.
3. Here I explain the nature and import of these patterns. The following section addresses the issue of their novelty, and hence their relevance to the epochal break thesis.
4. The point is that these authors acknowledged the importance of the problem of external validity for the natural sciences. A different question is whether their solutions are plausible. For a critical discussion of the latter question, see Radder 1996 (chaps. 4 and 6).
5. Note that this definition in principle allows for the occurrence of "noncommodified industrial research"—namely in those cases where industrial research is not dominated by (immediate) economic interests.
6. Concerning the mode-1/mode-2 approach, a recent review article (Hessels and Van Lente 2008, 754) concludes that "the empirical validity of the Mode 2 claims is limited."
7. See Weber 1949 and compare with Radder 1997. For an illuminating Weberian interpretation of Kuhn's philosophical theory of the history of science, see Mladenović 2007, 269–76.

REFERENCES

Bhaskar, Roy. 1978 [1975]. *A Realist Theory of Science.* Hassocks, England: Harvester Press.

Bok, Derek. 2003. *Universities in the Marketplace: The Commercialization of Higher Education.* Princeton: Princeton University Press.

Brown, James Robert. 2001. *Who Rules in Science? An Opinionated Guide to the Wars.* Cambridge: Harvard University Press.

Forman, Paul. 2007. "The Primacy of Science in Modernity, of Technology in Postmodernity, and of Ideology in the History of Technology." *History and Technology* 23 (1–2): 1–152.

Gibbons, Michael, Camille Limoges, Helga Nowotny, Simon Schwartzman, Peter Scott, and Martin Trow. 1994. *The New Production of Knowledge.* London: Sage.

Hamlyn, D. W. 1970. *The Theory of Knowledge.* London: Macmillan.

Hessels, Laurens K., and Harro van Lente. 2008. "Re-thinking New Knowledge Production: A Literature Review and a Research Agenda." *Research Policy* 37 (4): 740–60.

Krimsky, Sheldon. 2003. *Science in the Private Interest: Has the Lure of Profits Corrupted Biomedical Research?* Lanham, Md.: Rowman and Littlefield.

Kuhn, Thomas S. 1977. "Mathematical versus Experimental Traditions in the Development of Physical Science." In *The Essential Tension*, 31–65. Chicago: University of Chicago Press.

Latour, Bruno. 1983. "Give Me a Laboratory and I Will Raise the World." In *Science Observed*, edited by Karin D. Knorr-Cetina and Michael Mulkay, 141–70. London: Sage.

Mladenović, Bojana. 2007. "'Muckraking in History': The Role of History of Science in Kuhn's Philosophy." *Perspectives on Science* 15 (3): 261–94.

Petersen, Arthur C. 2006. *Simulating Nature: A Philosophical Study of Computer-Simulation Uncertainties and Their Role in Climate Science and Policy Advice.* Apeldoorn, Netherlands: Het Spinhuis.

Radder, Hans. 1996. *In and about the World.* Albany: State University of New York Press.

———. 1997. "Philosophy and History of Science: Beyond the Kuhnian Paradigm." *Studies in History and Philosophy of Science* 28 (4): 633–55.

———, ed. 2010. *The Commodification of Academic Research: Science and the Modern University.* Pittsburgh: University of Pittsburgh Press.

Shapin, Steven, and Simon Schaffer. 1985. *Leviathan and the Air-Pump: Hobbes, Boyle, and the Experimental Life.* Princeton: Princeton University Press.

Slaughter, Sheila, and Gary Rhoades. 2004. *Academic Capitalism and the New Economy: Markets, State, and Higher Education.* Baltimore: Johns Hopkins University Press.

Weber, Max. 1949. "'Objectivity' in Social Science and Social Policy." In *The Methodology of the Social Sciences*, translated and edited by Edward A. Shils and Henry A. Finch, 49–112. New York: Free Press.

Knowledge Making in Transition
On the Changing Contexts of Science and Technology

ANDREW JAMISON

THE MAKING OF KNOWLEDGE HAS BECOME an ever more integral part of our contemporary way of life. But much of the knowledge that is being made has little in common with what is usually referred to as "science." As social life has come to be infused with an overarching commercial mentality, science has lost much of its autonomy and the "academic freedom" that went with it. What was once a distinctly separate world of its own—a scientific community—has become a thing of the past, a figment of the imagination. "Looking for an expression that could capture the change that has occurred in the last century and a half in the relation between science and society, I can find no better way than to say that we have shifted from Science to Research," Bruno Latour (1998, 208) has written.

For Latour (1998, 208), "Science is certainty; Research is uncertainty. Science is supposed to be cold, straight and detached; Research is warm, involving and risky. Science puts an end to the vagaries of human disputes; Research fuels controversies by more controversies. Science produces objectivity by escaping as much as possible from the shackles of ideology, passions and emotions; Research feeds on all of those as so many handles to render familiar new objects of enquiry." It can be helpful to divide the changes that have taken place into two phases, one that was largely set in motion during World War II and came to a kind of climax in the 1960s, and the second from the 1970s on. While the first phase was primarily a massive scaling up of scientific activity, by means of vast increases in money and manpower, the second phase has brought about a fundamental change in meaning and operation.

From Little Science to Big Science

During World War II, scientists and engineers were supported by society to an unprecedented degree. They were given resources and opportunities to produce more effective weapons, from radar and rockets to atom bombs and toxic chemicals, as well as provide expert advice and secret intelligence that could be of value for the war effort. The mobilization of technology and science for war would initiate new kinds of relations between science, technology and society. From the 1940s through the 1960s, internally driven approaches to the production of knowledge, based on disciplinary identities and academic values, came to be complemented by externally imposed institutional forms and bureaucratic values. As Derek de Solla Price (1963) put it, when he summarized his statistical analysis of rates of increase in money and manpower devoted to research and development through the 1950s, "little science" had given way to "big science" (table 8.1).

In addition to the atomic bomb project, radar, computing, chemical warfare, and aircraft design and rocketry were major areas in which technology and science had been developed during World War II. And largely because of the decisive role that they seemed to have played in the war effort, the social status and prestige of science and technology changed significantly when the war had ended. The funding and organization of science and technology soon became a new area of concern for national governments, in addition to the relatively few

Table 8.1. Changing Modes of Knowledge Making

	Little Science Before World War II Mode-1	Big Science 1940s–1960s Mode-1½	Technoscience 1970s– Mode-2
Type of knowledge	disciplinary	multidisciplinary	transdisciplinary
Organizational form	research groups	R&D institutions	ad hoc projects
Dominant values	academic	bureaucratic	entrepreneurial

private corporations that had provided "external" support before the war. In the words of MIT engineering professor Vannevar Bush, who, after serving as a wartime government adviser, was asked by President Roosevelt to suggest how the U.S. government should best deal with this new role, science was characterized as the new "frontier." And the frontier that was science, unlike the frontier of the old West, was considered to be "endless."

The Bush report, *Science, the Endless Frontier*, discussed how the experiences of mobilizing science and technology for the purposes of war could be applied to peaceful purposes. On a discursive level Bush and his counterparts in other countries fashioned a strategic narrative: science was characterized as a crucial resource in what was soon to be perceived as an international power struggle. What had been achieved on the battlefields should now be achieved in the marketplace and in "international relations" (which became an academic subject of its own after the war). For strategic reasons substantially larger amounts of public funding ought to be channeled to science and education, or, as Bush (1945 [1960], 31) put it: "The Federal Government should accept new responsibilities for promoting the creation of new scientific knowledge and the development of scientific talent in our youth." In return for giving scientists and engineers vast amounts of money, Bush presented a glorious vision of unimagined prosperity and wealth based on applying science to the human condition, a sort of updated version of Francis Bacon's vision of *New Atlantis*.

At the institutional level a range of new research councils and other governmental bodies were created throughout the world; in the United States, Bush's report led to the establishment of the National Science Foundation. Major research and development (R&D) institutions were also built in the immediate aftermath of the war, particularly in order to develop "civilian uses" of atomic energy. These state-supported facilities complemented those already established by the military and provided a new set of large-scale, multidisciplinary sites for carrying out research and development activities that neither traditional universities nor private corporations could afford. These national laboratories came to be operated along industrial lines, and it was at one such institution—at Oak Ridge, Tennessee—that the director Alvin Weinberg coined the term "big science" as a way to distinguish the kind of knowledge produced at such places from the "little science" of the past (Weinberg 1967).

Combining traditional academic values with the demands of large-scale bureaucracies proved to be easier said than done, however, and there was a good deal of discussion as the 1950s progressed about the resultant cultural tensions—from C. P. Snow's famous lecture on the division of society into "two cultures" (a scientific-technical and a literary-artistic) to President Eisenhower's reflection on leaving office about the growing power of the "military-industrial complex." One of the main tensions concerned competing claims on the loyalty of the scientists. The so-called Oppenheimer affair of the early 1950s, during the anticommunist witch hunt, brought out some of the less attractive features of the new regime. J. Robert Oppenheimer, who had directed the Manhattan project (which had produced the atomic bomb), was stripped of his security clearance and his place on the Atomic Energy Commission (Salomon 1973). In the United States the new relations between science, technology, and society were challenged by such scientists as Albert Einstein and Leo Szilard and such critical intellectuals as Herbert Marcuse and Hannah Arendt, who had fled from Nazism, and saw the emerging order as a new form of authoritarianism (Jamison and Eyerman 1994). In Europe as well there were many scientists and philosophers who contended that the values of internationalism, academic freedom, and what Karl Popper more generally termed the "open society" were threatened by the new kinds of relationships that were developing between science, technology, and society (for a taste of the debate, see Shils 1968).

As the 1950s wore on, it became clear to politicians as well as to the general public that for all the money being spent on science, the promises of endless prosperity were still to a large extent unfulfilled. Science certainly contributed to ever more awesome weapons of mass destruction, but the Soviets seemed to be keeping pace. Now that Japan and the Federal Republic of Germany had been rebuilt and their economic structures reestablished, many American companies were facing intensified competition. In other "capitalist" countries, as well as in the Soviet Union, the state did not merely support basic research but applied technology and science to industrial development as well. The shock of the *Sputnik* satellite, which the Soviets sent into orbit in 1957, thus triggered important changes both in the discourses of science and technology policy as well as in the practical and institutional dimensions of knowledge production. For one thing it seemed to be insufficient to support scientific research and technological development without giving attention to the links between them—that is, how new scientific ideas were actually turned into new products. As economists began to explore the innovation process, as it started to be called, it became clear to many that technological innovations were not merely a matter of "applying" the results of "basic science," as Bush had implied in his report after the war, but required a more sophisticated understanding of firm strategies and the dynamics of technological development (Freeman 1974).

The changing relationships between science, technology, and society had a fundamental influence on the theory of science, as philosophers and historians came to debate the dynamics of scientific growth. On the one side were philosophers, led by Karl Popper, who contended that science grew continuously and cumulatively, and on the other side was the physicist-turned-historian Thomas Kuhn (1962), who recognized the social conditioning of scientific knowledge and presented science as a discontinuous process, a series of paradigm shifts and "scientific revolutions." Among economists and management consultants, there emerged a similar concern with processes of growth, and with the role of technical change in longer-term patterns of economic development. A particularly influential text of the early 1960s was *The Stages of Economic Growth* by W. W. Rostow, one of President Kennedy's advisers. Both scientific growth and economic growth can be considered central figures of thought in the dominant social and political discourses of the 1950s and 1960s. As the economist John Kenneth Galbraith (1968) came to characterize it, industrial society and its corporations were no longer seeking to maximize profit and produce goods and services for which there was a recognizable demand on an identifiable market; rather, modern corporations were in the business of producing technological development and pursuing growth as an integral part of the primarily military projects of the "new industrial state."

From Big Science to Technoscience

The various public debates and social movements that emerged in the 1960s served to challenge many of the assumptions of these discourses, and they contributed to opening science and technology to a range of new voices, constituencies, and concerns. Racial discrimination and ethnic integration, environmental protection and energy use, gender equality, and many other areas of research would become central topics of investigation in the years to come, both in university departments as well as in a range of new government agencies and research institutions.

In 1970 an Organization for Economic Cooperation and Development (OECD) committee, headed by Harvard engineering professor Harvey Brooks, produced the report *Science, Growth, and Society*. The report signaled the coming of a new era, in which the relationship between science, technology, and society would be substantially reconstituted. It was one of the most explicit attempts to transform the critical spirit of protest that was so strong in the late 1960s into constructive new kinds of policies. To facilitate these changes, there emerged a fresh conception of the state and of the exercise of political power—from representative government to what has come to be called governance (Mothe 2001). Rather than defining the task of the government primarily in terms of national security and military defense, the state took on a much broader role in relation to technology and science.

In the 1970s the view of a unified or universal science, based primarily on physics, which had dominated the theory of knowledge since the early nineteenth century, was challenged by what might be termed "pluralism": *Science* with a capital letter became a multiplicity of *sciences*. Within the natural sciences the "leading role" of physics was challenged by the rise of ever more mathematical and experimental "life sciences," and throughout the world the hegemony of the natural sciences was weakened by the emergence of new fields within the social and human sciences. Even more important, the dominant perception of Western science as the only legitimate form of knowledge production was questioned by various "ethnosciences" from other parts of the world as well as from minorities within the industrialized Western world (Jamison 1994).

The challenge to the hegemony of Western science and technology reflects the fact that in recent decades a handful of previously "developing" or undeveloped countries in Asia, especially China and India, have joined Japan as full-fledged competitors with the industrialized countries in many branches of industry, particularly in information and communication technologies. The so-called newly industrializing countries (NICs)—South Korea, Taiwan, Singapore, in particular—showed already in the 1970s that it was possible to develop successful export industries, not so much by developing links to science, along the lines of the growth and development storyline of the 1960s, but by focusing on particularly promising areas. As the Japanese had systematically tried to do, these NICs could "pick the winners" by means of technological forecasting and the creation of systems of innovation (Irvine and Martin 1984; Freeman 1987).

Such selective, or market-oriented, (re)industrialization strategies, which Western countries also began to foster in the 1980s, broke with the doctrines that had previously guided science and technology policy. They were not driven by security or military interests, but rather by purely economic concerns. With the coming into power of conservative governments in Britain and the United States around 1980, a new kind of commercial discourse entered the world of technology and science. Especially in the biomedical field, as well as in the broader areas of health and agriculture, powerful alliances would be created between transnational corporations and transnational organizations to reinvent knowledge in the name of the "life sciences."

The genetic breakthroughs of the 1950s, and in particular the double helix model of DNA constructed by James Watson and Francis Crick, served to trigger this process. The scientific-minded could use the code as a starting point for exploring the connections between particular kinds of genetic traits and particular kinds of diseases and plants, while the technically minded could try to build apparatus that could transfer genetic material from one organism to another. There was a clear potential relevance both for agriculture and medicine as well as for commercial applications. Some of the research carried out by university scientists in the 1960s was supported by companies interested in the results, but there were both bureaucratic rules and behavioral norms that served as barriers. In particular, the rights to what has come to be called "intellectual property" were complicated: should the universities where scientists worked be able to earn money on the results of the research, or were the scientists more to be regarded as the employees of the companies?

In the early 1970s, when scientists succeeded both in theory and practice in transferring genetic material from one organism to another, these issues took on an even greater significance. Already by the late 1970s, it was clear to many observers that genetic engineering, as it has come to be called, had enormous economic potential, and many of the scientists involved began to establish companies, where they could try to develop commercially viable products (Yoxen 1983). When Ronald Reagan was elected president in 1980, bringing with him an ideological distaste for bureaucracy (and for the "freedom" of scientists, as his tenure as governor of California in the 1960s had disclosed), such entrepreneurial activity received government encouragement, as did other kinds of efforts to strengthen the interaction between universities and industries. In the 1990s the international diffusion of the Internet, cellular telephones, and other information and communications technologies contributed to an intensification of contacts between universities and technology firms as well as to increasing attention to entrepreneurship and other aspects of knowledge management and product development.

Genetic engineering and information technology and, most recently, nanotechnology require expertise and skills from a number of scientific fields as well as an engineering competence, put together in what might be termed a "commercializable" cocktail. While certainly not all science has come to be integrated into processes of commercial innovation, there can be no denying that the rise of information technology and biotechnology industries has exerted a major influence on scientific research as a whole. As is readily apparent, these types of technology distinguish themselves from other types of technology in at least three major respects.

On the one hand, they are laboratory, or instrument-based, technologies, which means that they require major expenditures on scientific research and, most especially, expensive scientific instruments for their eventual development. And unlike the science-based innovations of the early twentieth century, which were for the most part applications of a scientific understanding of a particular aspect of nature (microbes, molecules, organisms, and so on), these new technologies are based on what Herbert Simon (1969) once called the sciences of the artificial. Information technology is based on scientific understanding of human-made computing machines, and biotechnology is based on scientific understanding of humanly modified organisms. Nanotechnology is the most recent example of a "mode-2" field that was based on the development of scientific instruments to make a previously unreachable realm of reality available for commercial product development.

Second, we are dealing with technologies that are generic in scope, which means that they have a wide range of potential applications in a number of different economic areas, social sectors, and cultural life-worlds. As opposed to earlier generic technologies, or radical innovations—the steam engine, electricity, and atomic energy, for example, which were primarily attempts to find solutions to identified problems—these new types of technologies tend to be solutions in search of problems. In this respect information technologies, biotechnologies, and nanotechnologies are idea-based rather than need-based, which means that in relation to their societal uses, they are supply-driven rather than demand-driven. That is one of the reasons why they require such large amounts of marketing and market research for their effective commercialization and indeed for their development.

Finally, these advanced or "high" technologies are transdisciplinary in what might be called their underlying knowledge base; that is, their successful transformation into marketable commodities requires knowledge and skills from a variety of different specialist fields of science and engineering. In earlier periods of technological development, there were clearer lines of demarcation between the specific types of competence and knowledge that were relevant; indeed, the classical categories of engineering are based on the particular types of scientific and technological theories that were utilized (chemical, mechanical, combustion, aerodynamic, and others). Genetic engineering and information technology and, most recently, nanotechnology require expertise and skills from a number of scientific fields as well as an engineering competence. The genetic engineer and the nanotechnologist certainly must know physics and chemistry and biology, but they do not know and learn these subjects in the same way as physicists, chemists, and biologists. Rather, they are taught to know what they need to know to provide the society with new sorts of products. As such, the novel technological fields represent a qualitatively fresh "mode" of knowledge production: a mixture of technology and science that has come to be called "technoscience."

For Michael Gibbons and his fellow authors of the influential book *The New Production of Knowledge* (1994), traditional discipline-based science—what they term mode-1—has been ever more supplanted by new forms of knowledge-production that disregard disciplinary boundaries and are directly oriented toward contexts of application. A new mode of knowledge production—so-called mode-2—is said to have emerged which challenges the values or norms that had previously governed the scientific enterprise. Knowledge in mode-2, we are told, has "its own distinct theoretical structures, research methods and modes of practice . . . which may not be locatable on the prevailing disciplinary map" (Gibbons et al. 1994, 168).

Under the influence of information and communication technologies as well as the challenge from Japanese firms, social and political theorists began to write in the 1980s about an "information society" that had come to be linked to the increasingly globalized character of commercial and financial transactions in the concept of the knowledge society. With the coming of the Internet and its enormous influence on the economy, as well as on the making of knowledge, Manuel Castells (1996) drew on many of these earlier efforts to provide an ambitious analysis of the "age of information" and the coming of what he terms the network society: "Toward the end of the second millennium of the Christian era several events of historical significance transformed the social landscape of human life. A technological revolution, centered around information technologies, began to reshape, at accelerated pace, the material basis of society. . . . *Our societies are increasingly structured around a bipolar opposition between the Net and the self*" (Castells 1996, 1 and 3, emphasis in original).

Like Latour and Gibbons and so many others who have tried to understand these changes, Castells can be accused of exaggeration. The emphasis on all that is new tends to downplay all that remains the same. There is a tendency toward hyperbole in such analyses. In certain respects they embody the developments they describe: academics going to market and scientists becoming the salesmen of the things that they create.

Hype, Habitus, or Hybrids?

In recent decades many if not most scientists have come to behave like businesspeople on the make, seeking out venture capital, lying about their results, and seeing each other as competitors rather than colleagues. Hyperbole, deception, and outright fraud have become an all too visible part of the research enterprise. These developments have led to a weakening—and in many places a complete elimination—of the sense of "enlightening" their fellow citizens that many makers of knowledge have traditionally thought they were doing. The idea or, perhaps more accurately, the myth of a pure, or basic science, driven by curiosity and the "internal" motivations of the scientific practitioners themselves and with an intrinsically beneficial societal function, has certainly not disappeared as an ideal, but it has become increasingly difficult to practice when the very possibility of carrying out scientific research has become dependent on the ever present need of funding and salesmanship. In the words of Derek Bok (2003, 207), the former president of Harvard: "Commercialization threatens to change the character of the university in ways that limit its freedom, sap its effectiveness, and lower its standing in society. . . . The problems come so gradually and silently that their link to commercialization may not even be perceived. Like individuals who experiment with drugs, therefore, campus officials may believe that they can proceed without serious risk."

Not only have the borders between science and business become increasingly blurred, and the meanings and practices of science increasingly commercialized. At the same time, the procedures of public accountability and the criteria of scientific legitimacy have changed character. Many is the scientist who goes directly to a company (or commercial media outlet) to sell his or her ideas, avoiding the traditional forms of peer review that have been a characteristic feature of the way in which the scientific enterprise has traditionally ensured "quality control."

Already in the 1980s, Aant Elzinga (1985, 207) had come to view the emerging commercialization with alarm, as he noted that established epistemic criteria—that is, the ways in which scientific truth claims are justified—were being put out of operation, as scientists increasingly found themselves in a condition of what he termed "epistemic drift": "the process whereby, under strong relevance pressure, researchers become more concerned with external legitimation *vis-à-vis* policy bureaucracies and funding agencies than with internal legitimation via the process of peer review. This may be seen as a process of erosion of the traditional system of reputational control."

In the ensuing years, that traditional system has been increasingly challenged by the dominant or hegemonic commercial culture. To borrow a term from the French sociologist Pierre Bourdieu (2004, 65), what might be characterized as the "habitus" of science—a mode of organization based in distinct scientific disciplines, which provides scientists with what Bourdieu calls a "collective capital of specialized methods and concepts"—has been ever more replaced by other modes or forms of organization. As a result, many scientists have reacted quite critically and have, like Bourdieu himself, sought to defend science from the intrusive commercial culture and other modes of organization. Many of the responses have, however, merely reaffirmed or reasserted the traditional beliefs or values of academic life, without recognizing that those beliefs or values have, to a large extent, become increasingly irrelevant. Like the national identities that are seen by neonationalist racists to be threatened by immigrants, the scientific identity is seen by many scientists to be threatened—both by commercialization and religious fundamentalism. But their attempts to reassert an autonomy or independence—like the attempts of the antiforeigners to reassert their national identity, which leave little place for immigrants to exist—tend to become reactionary and backward-looking and leave little room for creatively assimilating the new approaches to research or for dealing with the very real challenges, such as climate change, that we face as a species. In both cases, the response becomes a defense of a tradition, or a customary way of life, which to a large extent has outlived its usefulness and is no longer meaningful.

It can be helpful to consider these developments in relation to what Raymond Williams once termed "cultural formations." For Williams (1977, 133) social and cultural change involves the emergence of new "structures of feeling," new sensibilities, new mixtures of ideas and practices, or what Williams has termed "social experiences in solution." Emerging cultural formations or structures of feeling are subjected to two sorts of pressures, according to Williams: incorporation from what he termed the dominant cultural formations, and reaction from what he termed the residual cultural formations. In this sense we can think of the commercialization of research—or globalization—as a dominant cultural formation that seeks to incorporate all scientific and technical developments into processes of innovation; and we can think of the various forms of resistance or reaction—the defense of so-called Enlightenment values—as residual cultural formations, trying to adapt scientific and technological developments to older or more habitual ways of life and belief systems, to the traditional "habitus" or habitual behavior of scientists. Where the one attempts to commercialize all new ideas and inventions and turn them into business, the other seeks to uphold the old ideals of enlightenment and the disinterested pursuit of the truth that have traditionally been associated with big "s" Science.

It is between the traditional or residual regime and the commercial or dominant regime that an emerging or hybrid regime—a "mode-3" of change-oriented research is to be found (see Jamison 2001). It is by mixing and recombining elements of traditionalism and commercialism that hybrid knowledge makers carry out their various kinds of research. In terms of methods, organi-

Table 8.2. Research Cultures

	Traditional	Commercial	Hybrid
Research ideal	enlightenment	entertainment	empowerment
Main methods	analysis/ measurement	simulation/ codification	synthesis/ assessment
Organizational form	academic disciplines	competitive networks	cooperative alliances
Type of knowledge	theoretical/ objective	instrumental/ constructive	situated/collective
Form of education	disciplinary	professional	experiential

zational forms, types of knowledge, and approaches to education, the hybrid imagination represents a kind of synthesis of the commercial and the traditional systems or cultures of research (table 8.2).

The transition from science to technoscience is a long historical process that can be divided into two overlapping phases—a constructive or quantitative phase that started already in the nineteenth century and came to a kind of climax in the 1960s, and a more commercial or qualitative phase that has taken place since the 1970s. As a result, the traditional discipline-based approaches to science have not only become anachronistic but socially irresponsible as well. As the tiresome disputes about science and religion amply demonstrate, conducted in an air of mutual disdain and misunderstanding, science is only one way of knowing reality and provides knowledge about only certain aspects of existence. The contemporary resurgence of religious belief shows that the knowledge that is made by scientific methods, particularly as a commercialized mode of knowledge production has become ever more dominant, is simply not meaningful or helpful enough for many people. A hybrid or situated form of knowledge making based on experiential and collective learning processes and combining engagement and competence, theory and practice, professionalism and citizenship can perhaps help to reconnect science and technology to the needs and concerns of the broader society (Jamison 2008).

REFERENCES

Bok, Derek. 2003. *Universities in the Marketplace: The Commercialization of Higher Education.* Princeton: Princeton University Press.

Bourdieu, Pierre. 2004. *Science of Science and Reflexivity*, translated by Richard Nice. Cambridge: Polity.

Bush, Vannevar. 1945 [1960]. *Science, the Endless Frontier: A Report to the President for Postwar Scientific Research*. Washington, D.C.: National Science Foundation.

Castells, Manuel. 1996. *The Rise of the Network Society*. Oxford: Blackwell.

Elzinga, Aant. 1985. "Research, Bureaucracy, and the Drift of Epistemic Criteria." In *The University Research System*, edited by Björn Wittrock and Aant Elzinga, 191–220. Stockholm: Almqvist & Wiksell.

Freeman, Christopher. 1974. *The Economics of Industrial Innovation*. Harmondsworth: Penguin.

———. 1987. *Technology Policy and Economic Performance: Lessons from Japan*. London: Pinter.

Galbraith, John Kenneth. 1968. *The New Industrial State*. New York: New American Library.

Gibbons, Michael, Camille Limoges, Helga Nowotny, Simon Schwartzman, Peter Scott, and Martin Trow. 1994. *The New Production of Knowledge: The Dynamics of Science and Research in Contemporary Societies*. London: Sage.

Irvine, John, and Ben Martin. 1984. *Foresight in Science: Picking the Winners*. London: Frances Pinter.

Jamison, Andrew. 1994. "Western Science in Perspective and the Search for Alternatives." In *The Uncertain Quest: Science, Technology, and Development*, edited by Jean-Jacques Salomon et al. Tokyo: United Nations University.

———. 2001. *The Making of Green Knowledge: Environmental Politics and Cultural Transformation*. Cambridge: Cambridge University Press.

———. 2008. "To Foster a Hybrid Imagination: Science and the Humanities in a Commercial Age." *NTM—Zeitschrift für Geschichte der Wissenschaften, Technik, und Medizin* 1.

Jamison, Andrew, and Ron Eyerman. 1994. *Seeds of the Sixties*. Berkeley: University of California Press.

Kuhn, Thomas. 1962. *The Structure of Scientific Revolutions*. Chicago: University of Chicago Press.

Latour, Bruno. 1998. "From the World of Science to That of Research." *Science*, 10 April.

Mothe, John de la, ed. 2001. *Science, Technology, and Governance*. London: Continuum.

Price, Derek de Solla. 1963. *Little Science, Big Science*. New York: Columbia University Press.

Salomon, Jean-Jacques. 1973. *Science and Politics*. Cambridge: MIT Press.

Shils, Edward, ed. 1968. *Criteria for Scientific Development: Public Policy and National Goals*. Cambridge: MIT Press.

Simon, Herbert. 1969. *Sciences of the Artificial*. Cambridge: MIT Press.

Weinberg, Alvin. 1967. *Reflections on Big Science*. Cambridge: MIT Press.

Williams, Raymond. 1977. *Marxism and Literature*. New York: Oxford University Press.

Yoxen, Edward. 1983. *The Gene Business: Who Should Control Biotechnology?* London: Pan Books.

Alliances between Styles
A New Model for the Interaction between Science and Technology

CHUNGLIN KWA

BIOTECHNOLOGY AND NANOTECHNOLOGY have acquired, or almost acquired, a paradigm status of what science is today. Science is technoscience now, and philosophers of science are catching up with the recent status of technology vis-à-vis science. For better or for worse, the university is no longer the home of pure science. "If pure science ever existed," many would assert. But as Paul Forman (2007) reminded us, before 1980 neither scientists nor engineers, neither philosophers nor historians of science, and politicians the least of all, doubted the cultural primacy of science over technology.

If the cultural primacy of science was a myth, it was a myth with real consequences, not just in the twentieth century but in the early seventeenth century as well. At the very least, it contributed to forging a social relationship between science and technology, when Baconian science rallied with technology to improve upon human earthly existence. For this, it was enough that science should be useful, and this idea is older than the idea that science should lead to technological innovation. The very existence of a relationship between science and technology is of more importance than the question to which belongs primacy, every answer to which is always a reflection of the cultural standards of the age. Primacy has been taken to mean priority in discovery, but this it is not. The latter is an empirical, historical question to which no generalizable answer can be given. The once popular "linear model" of pure science leading to applied science leading to technology has justly been exposed not just as a myth but as wrong. It has never been true, but neither is or was its opposite. It is an extraordinary social fact that technology now enjoys cultural primacy over science. While this may have sensitized us to generally accepted examples of science developing out of a reflection on technology, and led us to the acknowledgment of the primacy of technology in several cases, this model is not generalizable either.

There is also a "model" of the science-technology relationship which simply denies that there is one. The idea of the autonomy of technology vis-à-vis science has been defended by a number of historians of technology, most recently by Thomas Misa (2004). The mere fact that a good history of technology can be written without reference to science is telling. Yet it could not say more than that the autonomy is relative and that there are analytical distinctions to be made between science and technology. A subtle yet pervasive mutual influence of science and technology in the seventeenth and eighteenth century has recently been demonstrated in *The Mindful Hand*, an edited collection (Roberts, Schaffer, and Dear 2007). The historians in this edited volume acknowledge a social distinction between natural philosophers and artisans, yet they also identify intermediaries, through whom skills and concepts were exchanged. Rich as their historical treatment is, from a systematic point of view the authors seem to do little more than vindicating Edgar Zilsel, who argued in 1942 that natural philosophers received their experimental skills from artisans and artists-engineers. What natural philosophers did with their newly acquired skills remains open for investigation.

But we could also read *The Mindful Hand* as saying that in concrete historical cases, alliances are forged between the various forms of thinking and practice of science and technology. The model I propose in this article is also a model of alliances, in which science and technology remain analytically distinct. But in addition, it breaks up science into six forms, six styles of science, in the way they were first proposed by Alistair Crombie in his 1994 grand overview of the history of science, *Styles of Scientific Thinking in the European Tradition* (Crombie 1994; see also Kwa 2011).

In a nutshell, Crombie's idea is as follows. In classical Greece an idea of science was developed that inaugurated a search for first principles. Ultimately, known phenomena should be derived with certainty from the first principles, hence deductive science. During the Renaissance several new styles of science developed: the experimental, the taxonomical, and a new form of theoretical science that deduced from hypotheses rather than from first principles: the analogical-hypothetical style. The deductive style, while remaining in place as an ideal, was restricted in practice to a few areas of mainly mathematics and a steadily decreasing number of topics in physics. During the nineteenth century two further styles came into being: the statistical and the historical-evolutionary style. Crombie obtained the number of six inductively, by historical investigation. There is no a priori reason why there could not be seven or eight, and perhaps there will be in the future. No style has as yet disappeared.

Each of the styles has brought together, in Crombie's words, "conceptions of the rational, the possible, the desirable, and the acceptable." Ian Hacking (1992) observed that each style is a "rather timeless canon of objectivity." Each style determines in its own way what qualifies as truth. Or falsity, as criteria for falsity are established along with criteria for truth, without which a certain kind of statement would not be recognized as a candidate for being "scientific" (Hacking 2000 [1982]). Without being exhaustive, logical certainty belongs to the criterion of truth of the deductive style, while the idea of reference of scientific concepts originated in the analogical-hypothetical style.

Crombie did not treat technology at all. Of the six styles of science, he gave separate treatments, thereby demonstrating their coming-into-being as in a sense "pure" styles, or at the very least the possibility to treat them as ideal-types. Nowhere did he put them in conjunction. Although to Crombie's credit it can be argued that the styles remained analytically distinct, later history of science abounds with examples of such conjunctions between styles.[1] In concrete cases the truth criteria of two, sometimes three, styles can operate at the same time and interact, but in each historical case the way they do so has to be solved ad hoc by the scientists who are involved. The conjunctions may therefore be seen as alliances, sometimes uneasy, sometimes more easygoing. An example of the latter is the alliance between the analogical-hypothetical style and the experimental style—an example of the former are the various ways in which statistical thinking is integrated into experimental practice. If we transpose this idea of alliances between styles of science to the science-technology relationship, we obtain in fact six theoretically possible varieties, perhaps more if we would allow for more styles of science to enter the equation. At any rate the way to think of science-technology relationships becomes enormously diversified.

But we have to first answer the question whether technology can be treated as style, on more or less equal footing as the six styles of science. Is there a single technological style? Are there perhaps styles, in plural? Distinctions such as science-is-knowing and technology-is-doing do not work, now that we know that much of science is doing, too, and that technology involves a good deal of knowing only part of which is derived from science. The concept of style in fact brings together ways of knowing and ways of doing. In the case of technology it should be able to identify what its "knowledge" aspect is.

Eugene Ferguson (1994) argues persuasively that design is *the* technological style. More precisely: *disegno*, or the engineer's drawing, by which a spatial arrangement of working components is achieved. Linear perspective, chiaroscuro, and the cutaway view were all Renaissance inventions, greatly enhancing the explanatory power of the drawing. The orthographic projection was developed by Gaspar Monge and James Watt in the late eighteenth century. The notebooks of Leonardo da Vinci and, among many others, Thomas Edison's three hundred years later, show that thinking by drawing has become the engineers' culturally entrenched habit. Ferguson relates how from the late 1950s on, instruction in engineering drawing began to disappear from engineering schools, by a dual effort of replacing drawing by computing and economizing on time spent on drawing. His own book played its part in restoring the balance toward the "art" aspect of engineering. We may also note that compared with the 1960s and 1970s, the computer changed roles and now enables visual thinking rather than suppressing it.

High Science, Low Science, and Technology

The cultural shift of about 1980 is more complicated than a shift from a primacy for science to a primacy for technology. The picture of science itself changed, too. Before 1980 one style of science stood for science as a whole: either deductivist science or the analogical-hypothetical style. Ernest Nagel's structure of science leaned toward the former. Popper's picture of science was the latter with strong deductive overtones, married to the experimental style of science that stood in a completely ancillary relationship to theory. In Kuhn's hands hypothetical reasoning became more loosely built on analogies, shown to be "world views," and pragmatically accepted as tools of discovery.

The hegemony of deductivist science contributed in no small part to the development of the linear model: it posited a logical relationship between science and technology. The precedence in time of science over technology would be more or less necessarily true on this account. Technology was a logical derivative of science, either on an analogy between experiment and a working technology (in Popper's view, if he would have discussed technology, which he did not), or by subsuming a working technology under a universal theory on the model of the reduction of experimental laws to theory (in the similar counterfactual case for Nagel). It was Mario Bunge (1966) who, leaning mostly on the Popperian approach, devised a philosophy of technology. Applied science, according to Bunge, is specifying the initial conditions so that a universal law becomes applicable to a desired process. Technology is using that process to some specified goal. The inclusion of styles of science such as the experimental style, but also taxonomy and statistics, into the picture of science contributes to make the Bunge model implausible.

Crombie published his work on styles of science in 1994, but his ideas had been circulating among historians and philosophers of science for decades. Ian Hacking, in his *Representing and Intervening* (1983), acknowledges Crombie's styles as a source of inspiration for his own successful attempt to liberate the experimental style from its subordinate position to theoretical science. Again, in his history of statistics Hacking (1990) asserts that statistics "fixes the sense of what it investigates," a statement that conveys adequately how styles of science set their own criteria of truth.

Hacking's work can be taken as a sign of a new interest in the "low" styles of science, the sciences that are dominated by data rather than by theory, and in which various forms of inductive reasoning are considered valid. Technology was not alone to profit from the loss of cultural legitimacy of "high" theoretical science. The taxonomical style of science is another case in point. Recently, taxonomy has received ample attention from cultural analysts such as Geoffrey Bowker (2005) and from historians of science. Ursula Klein and Wolfgang Lefèvre (2007) highlighted the importance of the taxonomy of chemical substances in mediating between the practice of artisans and speculation by natural philosophers. While historians discovered this for the eighteenth century, twenty-first century scientists of various denominations are putting taxonomy in the center of their work. The maps produced by the Human Genome Project (HGP), and by several other genomic and proteomic projects are ways of organizing data that are preeminently taxonomical. At the same time, the most salient characteristic of the HGP was the enormous effort put in developing gene-sequencing machines, producing the flood of data that were organized in maps. The HGP therefore may be considered as an example of an alliance between the taxonomical style and technology.

Technology and Experimentation

Hacking (1983, 230) argues that "to experiment is to create, produce, refine and stabilize phenomena." He mentions the photoelectric effect as an example of a phenomenon not existing anywhere in the universe before it was made in the laboratory. Hacking does not discuss the relationship between experiment and technology. While some of Hacking's examples come close to what is commonly designated as technology (mainly in the form of measuring devices), the context in which he discusses the specificity of experimentation is clearly science, not technology.

There is a similarity. Both the experiment and a working technology are interventions in the normal course of nature. The priority sequence between science and technology can therefore seemingly be reversed: experimental science is modeled on technical intervention. Modern science is an "uncanny manipulation of reality," wrote Martin Heidegger (1954, 52).[2] And twenty years later, Heidegger asserted that "science . . . is already the incessant incursion of technological representation into the realized" (qtd. in Forman 2007, 9).

Seen from the historical perspective of the adoption of the experimental method by science in the late sixteenth century, Heidegger was right. But we cannot conclude that therefore science *is* technology, even though experimentation and technology obviously have several features in common (intervention in nature, synthetic activity of the experimenter/ technologist), as Hans Radder (2003a and 2003b) has argued. The difference is not social uses, but rather that technology aims at a level of stabilization of phenomena that scientific experimentation does not need in the relatively disturbance-free laboratory. Moreover, scientists usually seek to "replicate" an experiment by different experiments (Radder 2003a and 2003b). The evaluation of whether the experiments deal with the same phenomenon will always refer to at least some theoretical notions.

Another point of communality between experimentation and technology may reside in the shared notion of design. But "design" in the case of experimentation is a way of speaking differently than conceived in technology: it refers to a "plan," like in "it was her design to become acquainted with him." Experimental design involves the skillful imposing of certain ceteris paribus conditions through certain actions, rather than originating from a drawing. Some famous experiments, like the Michelson-Morley experiment, do involve elaborate spatial setups, but very many, like in the biomedical sciences, don't. I believe therefore that we should see experimentation as one way of mediation between science and technology, rather than as a basis for equating them.

As I argued earlier in this chapter, we need to put the science-technology relationship on a broader footing than that between technology and experiment alone, even if it is the most salient bridge between technology and science. We surmise that there are possible relationships between technology and all of the other styles of science as well. The multifaceted analysis of aeronautical engineering by Walter Vincenti, in his *What Engineers Know and How They Know It* (1990), will serve as an entry.

A Second-Order Analysis of Vincenti's Case Studies

Vincenti's first case study is about the procedures of finding the best shape for the fore-and-aft section of the wing of an airplane, the shape of the air foil. Until the late 1930s a catalog of shapes were available for airplane designers. Wind-tunnel testing was the only selection procedure available, but usually far more shapes were on offer than could possibly be tested. Even then, to test two thousand shapes for a new airplane was not unusual. By 1945, however, air foil design had become "a logical, essentially rationalized process," according to Vincenti (1990), even when many uncertainties remain to this day (wind-tunnel testing has not disappeared). The original idea, leading to the new rational procedures, was to link air foil design to fluid mechanics, the intuition being that it would be advantageous to postpone the moment when the flow of air along the wing changed from laminar flow into turbulent flow. The intuition was successful though theoretically wrong, but the forged link between fluid mechanics and wing design was lasting. Fluid mechanics is now used to calculate air pressure distributions according to specific design requirements. The relationship between fluid mechanics and wing shapes is between a science that functions here according to the model of deductivism, even if incomplete.

Vincenti's second case study involves the establishment of consensus in the "flying-quality community" of what is a stable but not too stable airplane. Just like a bicycle, an airplane should be unstable because otherwise steering and correcting maneuvers would be too difficult; it should be stable because otherwise it would demand continuous supervision by the pilot, which is impracticable for any flight longer than a few hours. Please note the relationship of analogy between the bicycle and the airplane. For decades nobody had the faintest idea how to translate this subjective design requirement into a workable rule. During World War II an engineer (Robert Gilruth) in service of the (U.S. government's) National Advisory Committee of Aeronautics happened on the following "measure": "stick force per g," which makes the force a pilot should exert on the control stick a function of the airplane's acceleration. *Stick force per g* should be constant for one particular airplane, but it may vary across airplanes of different types, fighters and bombers typically being on extreme ends of the stability scale. Vincenti calls the ratio of stick force to acceleration "a rational criterion for manoeuverability." At any rate, once proposed, it was quickly adopted by the flying community.

This simple mathematical expression was not derived from any theory. Formally speaking, we are dealing with induction. It was proposed by someone who was familiar with lots of data on preferences by pilots but also had a good understanding of how an airplane behaves in flight, in particular when accelerating. Vincenti himself puts most emphasis on the long process of refining the specifications of "what pilots want" and selecting away theoretical considerations that proved of limited importance only. Gilruth's proposal (I avoid the term "discovery") thus appears to be based on "sophisticated theoretical reasoning," as Vincenti has it. But apparently, an intuitive selecting of which theoretical considerations were important among a wealth of others was at stake, and I propose therefore a similarity with Kepler's "discovery" of the laws of the planetary orbits in the wealth of data he had at his disposal. As Norwood Hanson (1958) made plausible, this was not pure induction. Something gestalt-like was involved. In principle, then, we are in the realm of the analogical style of reasoning.

Vincenti's third case study involves the way aeronautical engineers, and other engineers as well, incorporate theoretical considerations from thermodynamics and fluid mechanics into the design process. Vincenti notes that engineers work with the notion of a "control volume," volumes such as cylinders of steam engines through which fluids pass and on the borders of which thermodynamical relationships and processes can be specified. He observes that control volume is entirely absent in physics textbooks, in which control mass is used. Yet translations between the engineer's approach and the physicist's approach can be easily made, and there is a close proximity between the "first principles" of science as Vincenti calls them and the thinking practice of the engineer. Therefore, he argues, it would be a gross injustice to treat the control volume approach as a more applied version of the standard treatments of physics. Yet the control volume approach has a feature the control mass approach has not: a consideration of fluid flows. Thus we see thermodynamics put in a spatial mode of reasoning, typical for the design process.

Vincenti's fourth case centers on selecting the most efficient propeller blade for a given aircraft, from 1916 on, when systematic testing began. There were propellers with straight blades and with curved outlines, each in dozens of varieties of shape. No theory was available, hence testing was necessary, but testing was much more rational than simply cut-and-try. According to Vincenti, forms of experimentation were developed that were specific to engineering, not science, yet were not "applied science." The two outstanding features of propeller testing were the use of scale models and parameter variation. For a long time full-scale tests that aimed to identify the best propeller blade could be undertaken in flight only. Experimenting with scale models in wind tunnels was a practical solution, but the use of scale models requires "laws of similitude" to scale up results. The experiments themselves mainly revolved around parameter variation, the parameters being identifiable features of the blade design plus some conditions of blade operation. In a sense the design features served as the "theory" guiding the experiments.

In Vincenti's fifth case study we find technology "pure," without involvement of science except in rather elementary form and for self-evident reasons. It concerns the development of flush riveting. Protruding rivets are detrimental to smooth air flow and hence build resistance, causing speed reduction. Although this was recognized very early, no action was taken for a long time, since many other improvements promised greater gains. When these were achieved, however, by 1938, virtually all the aircraft companies in the United States set out to get rid of the protruding rivets at the same time. The central design problem was finding the optimum angle of the rivet head. Joining thin sheets of metal apparently required wide-angle rivet heads, lest the sheets would crack. The technique of hammering mattered, one blow giving better results than several softer blows. An artisanal experience with materials therefore was indispensable for solving the design problem. In sum, several aspects of technological practice worked together in the case.

Vincenti defines design as the content of a set of plans (usually in the form of drawings) and the process by which those plans are produced. He further analyzes the various aspects of the design process as follows: fundamental design concepts, criteria and specifications, theoretical tools, quantitative data, practical considerations, and design instrumentalities. On the basis of his own case studies, I have focused somewhat more narrowly on the spatial thinking embodied in drawings. Yet this spatial thinking is more integrative than the pure geometrical features of the drawing by itself: I assume, for instance, that no engineer can contemplate a drawing without thinking of the materials that are involved and hence the doability of putting them together.

Vincenti also offers a number of categories on how to think the way science and engineering interact mutually: transfer from science, invention, theoretical engineering research, design practice, and so on. These I have replaced by the simpler scheme of alliances between design and technology and various other styles of science, different ones in each case.

Although the idea of alliances between technology and a scientific style recognizes the specificity of technology as a style in its own right, it goes against the

claim that technology is "autonomous," at least in very many cases of technological development. The model of alliances between styles implies that neither
technology can be subsumed under science, nor vice versa, and this is entirely
in keeping with Vincenti's own stance. As we have seen, there is in fact not one
alliance between technology and science but several: six styles of science allowing for the possibility of, in principle, six different alliances and possibly more
if multiple styles of science are involved in more than one alliance. The model
of alliances also allows us to qualify Paul Forman's (2007) thesis of a cultural
shift, a reversal of cultural primacy with technology now leading at the expense
of science. Even when technology is the "leading partner" in an alliance with
science, we would not expect a total eclipse on the part of science.

But there is another side to this as well. Although the shift from a cultural
primacy from science toward technology may be experienced as dramatic, an
exclusive focus on this shift alone would blind us to an equally dramatic shift
that operated within science itself. This is the shift in dominance from the
"high" styles to the "low," from a focus on theory to one on data and ways of
producing data. It may be expected that both shifts reinforce one another. The
data-producing styles in science are dependent on various forms of technology
to produce and handle data. Yet alliances may not be stable over long periods
of time. The relationships between science and technology, and within science
itself, are too diverse to expect that one single configuration will be dominant
for a long time to come.

NOTES

1. See also Hacking 1992. Hacking suggests that styles combine as "mixtures" rather
   than as "compounds" in which case there would be a new style.
2. From Heidegger's original German: "unheimlich eingreifende Bearbeitung des
   Wirklichen" (Heidegger 1954, 52).

REFERENCES

Bowker, Geoffrey. 2005. *Memory Practices in the Sciences*. Cambridge: MIT Press.
Bunge, Mario. 1966. "Technology as Applied Science." *Technology and Culture* 7:
   329–47.
Crombie, Alistair C. 1994. *Styles of Scientific Thinking in the European Tradition: The
   History of Argument and Explanation Especially in the Mathematical and Biomedical
   Sciences and Arts*, 3 vols. London: Duckworth.
Ferguson, Eugene. 1994. *Engineering and the Mind's Eye*. Cambridge: MIT Press.
Forman, Paul. 2007. "The Primacy of Science in Modernity, of Technology in
   Postmodernity, and of Ideology in the History of Technology." *History and
   Technology* 23: 1–152.

Hacking, Ian. 1983. *Representing and Intervening: Introductory Topics in the Philosophy of Natural Science*. Cambridge: Cambridge University Press.

———. 1990. *The Taming of Chance*. Cambridge: Cambridge University Press.

———. 1992. "'Style' for Historians and Philosophers." *Studies in the History and Philosophy of Science* 23: 1–20.

———. 2000 [1982]. "Language, Truth, and Reason." In *Historical Ontology*, 159–77. Cambridge: Harvard University Press.

Hanson, Norwood R. 1958. *Patterns of Discovery: An Inquiry into the Conceptual Foundations of Science*. Cambridge: Cambridge University Press.

Heidegger, Martin. 1954. "Wissenschaft und Besinnung." In *Vorträge und Aufsätze*, 41–66. Pfullingen, Germany: Neske.

Klein, Ursula, and Wolfgang Lefèvre. 2007. *Materials in Eighteenth Century Science: A Historical Ontology*. Cambridge: MIT Press.

Kwa, Chunglin. 2011. *Styles of Knowing: A New History of Science from Ancient Times to the Present*. Pittsburgh: University of Pittsburgh Press.

Misa, Thomas. 2004. *Leonardo to the Internet: Technology and Culture from the Renaissance to the Present*. Baltimore: Johns Hopkins University Press.

Radder, Hans. 2003a. "Technology and Theory in Experimental Science." In *The Philosophy of Scientific Experimentation*, edited by Hans Radder, 152–73. Pittsburgh: University of Pittsburgh Press.

———. 2003b. "Toward a More Developed Philosophy of Scientific Experimentation." In *The Philosophy of Scientific Experimentation*, edited by Hans Radder, 1–18. Pittsburgh: University of Pittsburgh Press.

Roberts, Lissa, Simon Schaffer, and Peter Dear, eds. 2007. *The Mindful Hand: Inquiry and Invention from the Late Renaissance to Early Industrialisation*. Amsterdam: Koninklijke Nederlandse Akademie van Wetenschappen.

Vincenti, Walter. 1990. *What Engineers Know and How They Know It: Analytical Studies from Aeronautical History*. Baltimore: Johns Hopkins University Press.

PART II

Experimenting with the Concept of Experiment
Probing the Epochal Break

ASTRID SCHWARZ and
WOLFGANG KROHN

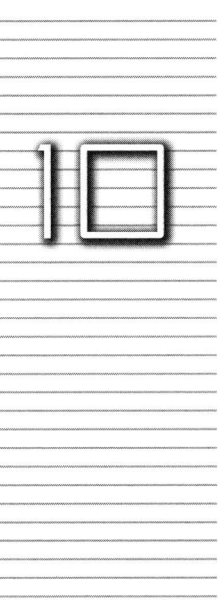

FOR A COUPLE OF YEARS NOW a chorus of rather cacophonic voices has been heralding the fact that over the past few decades science has undergone a profound transformation. This has been answered by another chorus, more precise and concordant, that there has been no such transformation—at least no break or sharp discontinuity—and that the existing, commonly accepted vocabulary is sufficiently apt to describe recent developments in science and society. Beyond the parameters of this so far indissoluble antinomy, several voices have been attempting to overcome this dualistic formation. In this chapter we focus on one such position that offers a specific mode of analysis for identifying such a transformation, for "seeing" an epochal break. Our intention in using this term is not to question whether or not knowledge production as a whole has changed over the past few decades. Instead, we adopt the position that it is rather a question of finding the right vantage point—that is, "the proper distance to scientific practice"—of making a case rather than settling a fact (see Alfred Nordmann's chapter in this volume). Such a vantage point enables us both to make visible and to appreciate the changes that might turn out to be important and illuminating in better understanding the forces and powers governing relationships between science, technology, and society.

Our focus is on observing changes in the practices of science-based experimentation in society. We suggest that there has been a major shift from the *laboratory ideal* to the *field ideal* of experimentation. The laboratory ideal involves designing manipulated, well-controlled, isolated experimental systems; the field ideal acknowledges their complexity, blurred boundaries, and unpredictable response to interventions. Field experiments could hardly be called an alternative ideal if they had not undergone a reevaluation in the philosophy of science and a reassessment with regard to their social relevance. We suggest that both changes can be observed especially well in the 1980s in the domain of the environmental sciences. Even if field experiments were not entirely new at that time, environmental concerns in science and society gave them a new cognitive status, institutional backing, and a specific rhetorical image. Today we are seeing the spread of new styles of experimentation to many areas of society. Experiments performed in *open spaces* might be, say, a social reform or a medical treatment, an ecological remediation or a technological innovation. A number of concepts are in circulation that seek to label these various experimental constellations, including real-life or real-world experiments, experimental installations or innovations, adaptive or experimental management, and prototyping. For the sake of conceptual clarity, we have decided to use "field experiment" as a generic term.[1]

The dynamic interaction between research activities and innovation strategies forms an important feature of what is widely known as the knowledge society. One of its outstanding features is the continuous shifting of knowledge-production into contexts of application and a concomitant increase in research in the applied sciences. This trend not only signals the growing relevance of applied knowledge in all domains of society; it also implies the extension of research practices to sites outside the institutional framework of science. Furthermore, if research—both basic and applied—comprises experimentation, then clearly experimental activities can be expected increasingly to pervade every field of innovation in society. In the process the institutional rationality of science that welcomes errors and failures as vehicles for augmenting and substantiating knowledge is transferred to society, at least to some degree. Society in turn confronts science with new responsibilities regarding the risks associated with research in the open spaces of societal change.

Scholars of the philosophy of science have generally paid little attention to these changes. A brief glance at recent literature may help to position our argument. Martin Carrier proposes an "interactive view" of the relationship between science and technology. He argues: "For letting this potential of reciprocal stimulation unfold[, it is essential] to leave room or leisure for hooking up the practical goals with the theoretical framework" (Carrier 2007). This evokes a concept of research that comes close to the Baconian notion of the ideal mode of interaction between science and society: maintaining an awareness of epistemic challenges as they emerge alongside societal needs and ensuring that they are reflected in practical research goals. We will return to the Baconian conception in a moment, but let us first hear another voice. In his 1997 book *Pasteur's Quadrant*, Donald Stokes argues against a linear, unidirectional model of knowledge production and transfer from basic-pure science to applied science and technology. Instead, he suggests that technological advances can also lead to a deeper understanding of a particular theory. He puts forward a four-field scheme based on the parameters of "understanding" and "control," resulting in four different research modes that are all equally present in the world of knowledge production. One such research mode—"Pasteur's Quadrant"—stands for use-inspired basic research, and it is this approach on which the author focuses especially and toward which his sympathies are clearly directed, given that Stokes has chosen it as the title of his book. Although researchers in this quadrant are fully aware of the potential real-world utility of their work, they never lose sight completely of their desire to advance scientific understanding as well. *Pasteur's Quadrant* calls to mind the Baconian ideal—an impression strongly supported when it comes to Stokes's ideal of a symbiotic relationship between science and government in the service of human welfare (figure 10.1).

However, we do not believe that the Baconian ideal of relating science to society—the so-called Baconian contract—is capable of addressing the spread of scientific practices into innovative fields of society. Instead, we are heading toward a new knowledge-society contract, which on the one hand turns society

| | | Interest in application | |
		No	Yes
Seeking Basic Understanding	Yes	"Bohr" pure basic research	"Pasteur" use-inspired basic research
	No	?	"Edison" pure applied research

Figure 10.1. Four-field schema in **Pasteur's Quadrant**. *Source*: Stokes 1997.

into a research field and open laboratory and on the other binds experimental practices and hypothetical reasoning to the conditionality of social acceptance. We regard this shift from the Baconian to the knowledge-society contract as the pivotal point of the epochal break.

So are we really heading toward a new knowledge-society contract? It could well be argued that the features of field experimentation are not entirely new. Indeed, history abounds with examples from medical research, weapons testing, schooling and learning, colonialism, and the chemical industry (to name just a few) where human beings have been unduly exposed to strategies of trial and error. This kind of "science in the making" has been observed in authoritarian as well as democratic societies. However, in only a few cases have those concerned and the general public been properly informed—let alone invited to participate actively. Usually, the experimental design involved was both secretive and sloppy. Given this backdrop, this kind of experimentation stands on ethically and politically slippery ground. In line with the self-image of science, however, historical cases of problematic field experimentation are seen merely as blots on an otherwise unblemished landscape of laboratory research. If it is true that scientific research is increasingly becoming an agent of change, then new forms of legitimacy, information, and participation are needed that can be readily interpreted as giving rise to a new knowledge-society contract.

1. Philosophy's Blind Spot Regarding Experimentation

Experimentation became a focus of interest for philosophers of science from the 1980s on, when they began to pay attention to the role of scientific practices in knowledge building rather than dealing solely with theories and purely logical operations.[2] A whole range of new research questions arose as a result, giving rise to the new research field of experimentalization.[3] Since then, it has become commonly accepted that there is more to experimentation than just the testing of theories. There is now notably less agreement around the question of how experiments address theories and even less about whether theory always plays a role in experimental practice.[4] Nonetheless, these activities aimed at revisiting the role of scientific experimentation ("the new experimentalism") rarely touched on the field ideal of experimental practice. One reason for this may be that the field ideal is even more of a pluralistic concept than the laboratory ideal. It can be located in such diverse disciplines as psychology, sociology, geology, and economics. However, there has been no broader systematic attempt to date by the history and philosophy of science community to compare and scrutinize these different approaches.

Despite this rather unsatisfactory epistemological situation, we have identified an important locus of theoretical activity that crystallized in the 1980s and was prompted by the environmental movement. One move that brought about profound change—namely, the search for a different cognitive structure of scientific knowledge—was the call for ecology to be an "alternative" or "soft" science.[5] Another was the growing awareness of the social and political context of knowledge production. "There is a social interest at stake not only in the utilization of scientific knowledge but also in its production" (Böhme 1979, 105). A research group comprising philosophers and sociologists made its main concern the analysis of "alternatives" produced by science that were later "filtered out" by dominant societal interests. One of the outcomes to emerge from this project in the 1970s was that ecology was now regarded as a scientific field of knowledge production that might enable us to understand ourselves not only as a product but also as an agent of nature. The frequent reference to "society as a laboratory" captures quite well the ongoing transformations that coalesced around a concept of experiment that we have called the "field ideal."[6]

Perhaps the most striking feature of field experiments is that they deal with objects "outside," in an uncontrolled environment. Further important features in the field include individuality, uniqueness, contingency, instability, and also potentially lack of safety. This kind of conceptualization clearly establishes a distinction between the individualizing, value-laden understanding of field objects and that of lab objects as instances of generalizable knowledge. However, laboratory practice is no substitute for field experiments: "unforeseen difficulties are found only in pilot experiments conducted under field conditions."[7] It is the field ideal of experimentation along with its historical, epistemic, and rhetorical tradition that becomes the dominant framework within which scientific findings and technological innovations in our knowledge society are first tested and then eventually applied. The knowledge society itself—and not the controlled spaces of a laboratory, a museum, a court, or a theater—is the stage upon which experimental design must prove itself.

2. Historical Flashback: The Baconian Contract

It was not until the late seventeenth century, after the founding of the new academies and societies of modern science, that the laboratory ideal of experimentation was codified. Its central dogma is that scientific experimentation—whether it results in success or failure—cannot cause harm to society as long as it is performed within the walls of these institutions. These walls facilitate an unlimited search for facts and the construction of artifacts for the purposes of expanding and testing the knowledge base of science. In this sense the meaning of "walls" can be taken quite literally. Taken metaphorically, they hint at a structural analogy between the methodological isolation of experimental systems within their natural environments and the ideological isolation of experimental activities within their social environments. However, as a new style of innovative practice, the experimental spirit was not created within "walls" but spread out into many fields of late Renaissance society. Artists, engineers, instrument makers, surgeons, and other practitioners developed new attitudes toward understanding nature's inventions, designing machinery, and exploring the globe.

Francis Bacon gave a variety of names to this new form of intellectual activity: "experimental philosophy," "scientia operativa," and "inquisition of nature." He also articulated what came to be called the "Baconian contract" between modern science and society. If the normative structure of society is prepared to permit all kinds of investigation into the causal structure of nature—the nature of human beings and society included—then science is prepared to pass back potentially useful knowledge and technology to all spheres of society. Bacon's lifelong (albeit unsuccessful) efforts to gain political support for organizing experimental research on a large scale caused him to ponder the question, What kind of institutional setting would convince society of its benefits? Because the promise of gains cannot be justified by an anticipatory form of argument, he suggested in *Novum Organum* that balancing social costs and benefits was a matter of risk and trust: "For there is no comparison between that which we may lose by not trying and by not succeeding; since by not trying we throw away the chance of an immense good; by not succeeding we only incur the loss of a little human labor. . . . It appears to me . . . that there is hope enough . . . not only to make a bold man try [ad experiendum], but also to make a sober-minded and wise man believe" (Bacon 1860 [1962], book 1, aphorism 114).

This assessment of the risks associated with the political authorization of the experimental method was based on an important normative claim concerning the relationship between science and society—namely, experimental failures as well as errors of hypothetical reasoning are acceptable because they affect only the internal discourse of science, not its social environment. Mistakes in the laboratory can easily be corrected, and society is only affected in terms of its choice of options from among those offered by approved scientific knowledge. These conditions applied to experimental science have served as the backbone of the dominant ideology that supports scientific progress, making scientific research and technological invention key features of the process of organizing and modernizing society and its institutions. The Baconian conception of experimental science became the foundational element in the contract between science and society (Gibbons et al. 1994; and Schäfer 1999) and between society and nature (Serres 2000).

3. Laboratories as a Strategy for Generating Failure without Failing

Laboratories are protected spaces. Their isolation serves to reproduce, standardize, and generalize experimental findings. Stripped of contextual complexity and environmental variation, they are breeding zones for strange, unforeseen, and unapproved knowledge, skills, and techniques. In a metaphorical sense the isolated laboratory is also a precautionary principle that protects society from experimental failures and errors of hypothetical reasoning. The laboratory is the only place in society where failures and errors are welcomed, respected, and even morally valorized. In any case, so the reasoning goes, society is free—and has the responsibility—either to adopt and apply scientific knowledge or else to reject it.

As the dominant mode of legitimizing scientific research, the Baconian contract prevailed throughout the nineteenth and most of the twentieth century. It remains powerful even today. However, during the course of the mutual development of science and society, experimentation became a polymorphous concept and its social relations multiplied. On the one hand, the ideal of laboratory experimentation became more rigid in terms of reproducibility and precision as well as in its function to serve theory formation. On the other hand, it was modified as it spread to include all kinds of objects—nonliving and living, psychological and social, natural and technical, simple and complex, constant and changing, very small and very large, frequent and unique, within well-defined and ill-defined boundaries, with well-controlled parameters and uncontrolled field conditions. In the industrialization process of the nineteenth century, scientific experimentation became linked with experimental practices of innovation in various societal sectors, such as in agriculture (testing Liebig's artificial fertilizers) and in the health sector (e.g., in the vaccination campaigns of Pasteur and Koch).

Ecological field experiments also contributed to a shifting of the boundary: at the beginning of the twentieth century, entire lakes were used to perform experiments aimed at finding out more about the lacustrine nutrient cycle. The Schleinsee, a small lake in southern Germany, became a famous experimental system, enabling some of the most important issues concerning the complex phosphate cycle to be clarified.[8] For this purpose the whole lake was artificially fertilized with phosphates. Today this experiment would be designed to include many more actors, such as conservationists, residents, water sports enthusiasts, anglers, and—if we are prepared to include Latour's nonhuman agents—ducks and water fleas, among others. It would probably be rejected on legal, political, and ethical grounds.

Even if historical justice demands that we pay attention to these varied modes of scientific experimentation performed in open fields, laboratory experimentation has remained the ideal type of knowledge acquisition. This is especially true of all research activities and new technologies that imply risks to life and health. However, it was illusionary from the beginning to set limits to the spread of theoretical ideas that might imply risks for minds and morals. As the most prominent example—Darwinian evolutionary theory—shows, ethical neutrality and acknowledgment of error do not count for much on either the proponents' or the opponents' side when strange and novel ideas become issues of public discourse. Indeed, the principle of freedom of opinion in democratic societies ensures—with a decreasing number of exceptions—that scholars can make public and even argue forcibly for the application of theories that are still in the making. The premature testing of half-baked theories has been a frequent feature of this trend. More seriously, science- and technology-based attempts to participate actively in modernizing society have become more and more successful and have left traces in many arenas. And yet the legitimacy of these attempts was largely provided and protected by the Baconian contract, whose elasticity is gradually being stretched to the point of exhaustion. A new formula capable of handling the interaction between research and innovation is called for—and is even being practiced in certain fields. Talk of an epochal break should not be mistaken as signifying a sudden change of structure but rather seen as offering a new semantics for an ongoing process of change.

4. Social Experiments with Society

Even this semantics has its precursors. John Dewey (1929, 133) was a prominent proponent of the idea that experimental knowledge production and social change are interwoven: "The ultimate objects of science," he wrote, "are guided processes of change," and truths are "processes of change so directed that they achieve an intended consummation." Certainty in knowledge follows from achieving reliability in action. Even more radical is the notion of societal experimentation developed by the Chicago school of sociology. Albion Small (1921, 187) held the view that the rapid change of modern settlements in itself provides a "world of experimentation open to the observation of social science. The radical difference is that the laboratory scientists can arrange their own experiments while we social scientists for the most part have our experiments arranged for us."

Small located the idea of experimentation in social life and not in the scientific method. This notion of experimentation became influential in American sociology. After World War II, Karl Popper—struggling with totalitarian political experiments—suggested "piecemeal social engineering" as a way to introduce scientific method into politics. In 1969, Donald Campbell wrote the influential article "Reforms as Experiments." He developed a methodology comprising political planning and scientific design of social experiments. Although objections have been raised regarding the technocratic attitude of these approaches (i.e., that reforms were imposed more or less coercively on people), they have been influential in more recent attempts in which those affected have been turned into participant observers. These various new approaches and concepts can, with hindsight, be seen as foreshadowing the search for a new contract between science and society. But they can be seen neither epistemologically nor politically as a coherent line of development.

5. Scientific Experiments in Society

We now return to our main point, the epistemological shift from laboratory to field experimentation. The differences can either be emphasized by contrasting ideal cases or they can be interpolated by pointing at cases in which aspects of both are combined. We will give a rather rough-and-ready outline of the ideal-types and then focus on the "dappled world" of field experiments, which offer features of both. Field sciences search for and find their objects "outside," in an uncontrolled environment. Nevertheless, they also perform experiments, and these experiments are necessary; they cannot be replaced by laboratory practice. Selecting, reading, modifying, and comparing places are essential elements of field practice. Natural places are not just neutral stages on which scientific activities are played out, as labs are; rather, they are themselves objects of study. Plants and animals and, of course, human agents are not more or less passive "guests." The field experiment is based on a different material setting and metaphysical understanding than the lab experiment.

However, perhaps the most important feature is that field experiments are done in and with particular and variable places, and that each of these places is the result of a particular and unique history. Historian Robert Kohler (2002, 6) has described the field sciences as being mainly practices of place: "Field biologists use places actively in their work as tools; they do not just work *in* a place, as lab biologists do, but *on* it. Places are as much the object of their works as the creatures that live in them." To sum up, we might say that the field ideal of experimentation is oriented toward a practice of place, where spatial openness, individuality and uniqueness, instability and contingency are the dominant features. Unsurprisingly, this stands in sharp contrast to the conceptualization of the laboratory ideal, which is characterized by isolation, intervention, and completeness (Carrier 2006, 21).

We next present three types of experiments, each of them characterized by a different constellation of field and lab elements. On the one hand, the three types give an idea of the wide range of possibilities for recombining the features of the field and the lab ideal. On the other hand, they help clarify—systematically though not historically—the argument we have put forward about the shift in the concept of experiment. We start with an example of a lab situation aimed explicitly at simulating the field—not digitally with computer technology, but using an extracted segment of the "real" situation, which is why we call it real-world simulation. In the second example we learn something about nature's experiment, probably the most pure design of a field experiment. Then we turn to a case where scientific research is itself becoming an agent of change in society, thereby giving rise to a new knowledge-society contract—an epochal break.

5.1. Tank in the Lab, the Real-World Simulation

The Max Planck Institute for Limnology in Plön (northern Germany) had installed two so-called plankton towers in its laboratory—"pillars to science" as they called them.[9] The plankton towers were two steel pipes 12 meters in height and 85 centimeters in diameter, filled with around 10,000 cubic meters of lake water. Every 50 centimeters the column was equipped with a set of sensors for measuring and controlling temperature, pH, and light in situ. Interventions in these water packages were possible by injecting chemicals or algae and by extracting "lake water." In this comprehensively controlled water column, experiments were performed either with particular plankton species or with plankton communities. The institute's website stated programmatically that the plankton towers are intended to fill the gap between lab experiments and field experiments. At the same time, researchers state that this is also a source of the problem they face when attempting to transfer the facts discovered in the towers to the field situation in the lake.

What we learn from this case, which is oriented more toward the lab than the field, is that it is precisely the elements of instability and contingency that make it necessary to control the environmental conditions as well as to control the space to perform these experiments. At the same time, these experiments gain their epistemic value precisely because they are related to the field in this way—even if the character of this relationship cannot be fully understood. Karin

Figure 10.2. Field site at the Center of Advanced Studies (Zentrum für interdisziplinäre Forschung, ZiF) in Bielefeld, Germany. Studies about the invasive species *Heracleum mantegazzianum.* Photo courtesy of Astrid Schwarz, June 2007.

Knorr-Cetina (1999) speaks of a gap that will never disappear completely even if the laboratory allows for an improvement (in the sense of better understanding and control) in the relationship between the natural and the social order. One might say that this kind of laboratory experiment constructs and simulates the phenomenon whereas field experiments construe and prompt the emergence of something rather more unruly than a "phenomenon" (figure 10.2).

5.2. Invasive Environment and Species—Nature's Experiment

A red- and white-striped ribbon is all that protects this scientific object—a group of plants, the species *Heracleum mantegazzianum*—from its invasive environment in the grounds of the Center of Advanced Studies (Zentrum für interdisziplinäre Forschung, ZiF) in Bielefeld. The milieu of the plants is dominated by wind and precipitation events and disturbed by curious deer, humans, falling branches, and in some cases aggressive conservationists waging a struggle against this alien species. In addition to the perceived "invasiveness" emanating from their surroundings, these objects themselves are seen as being "invasive."[10] These species are so-called alien species and are identified with the problem of biological invasions, which has attracted increasing attention in ecological investigations. The resulting literature is impressive, as is the variety of statements made to explain why this research is important. The following quotation from the website of the Institute of Ecosystem Biology at the University Bielefeld is representative of such statements: "Biological invasions represent great natural experiments for the ecologist whose investigation is extremely valuable for the understanding of population spread and community- and landscape-level processes affecting the patterns and abundance of species at large spatial and temporal scales, that is, scales which are otherwise hardly accessible for experimental ecologists."[11] This strip of meadow with the invasive plants is a virtually uncontrolled place where (most) instabilities and contingencies are welcomed because they are the objects of scientific investigation. The special quality of the experiment consists in the minimal invasiveness of the experimenter, which enables the agency of nature to come to the fore.

5.3. Piloting Innovation in Society

The restoration of a lake is a typical innovation experiment. A detailed reconstruction of the restoration of Lake Sempach in Switzerland showed how the specific conditions of a lake point to the necessity to study its individuality in the context of a real-world experiment (Groß, Hoffmann-Reim, and Krohn 2005). Still, scientists expect that despite this individuality, the results might in some respects be generalizable and transferable to other restoration projects. Consequently, the knowledge acquired in such projects often results in a form of expertise where experience gathered by observing particular cases merges with scientific background knowledge. Generally, these kinds of experiments are performed in the context of innovation projects. They are supported by the idea that social and technological innovation not only call for scientific experimentation to be extended outside the laboratory, but also that all relevant aspects of society and nature become involved. The restrictive and protected closed space of the laboratory is left behind. Instead, experimental devices are brought "outdoors." Test stations, prototypes, pilot installations, ecological restoration projects, test releases of drugs, pedagogical reform projects, town district developments, and so on expose scientific knowledge to unrestricted reality conditions. At the same time, projects of this kind demand planning, monitoring, data processing, and interpretation. Features of this type of experimentation aim at identifying and taming surprises, which would rarely appear in the laboratory world. Innovation experiments intend "to turn the relationship between action and surprise into an experimental design" (Krohn and van den Daele 1998, 195).

6. Epochal Break Seen through the Lens of the "Experiment" Concept

The importance of science in society derives not only from its ever increasing contribution of knowledge but is based above all on the transfer of experimental practices to the design, monitoring, and evaluation of innovation processes. The search for new sources of energy is a good example. The development of potential scenarios is unavoidably based on assumptions regarding energy resources, new technologies, and consumer life styles that are in turn heavily dependent on recent scientific knowledge or even hypothetical reasoning. Thus decisions based on scenarios necessarily contain experimental elements. Monitoring and feedback mechanisms determine the development of novel strategies that are either reinforced by success or weakened by surprise. Even if there is an element of path-dependent lock-in arising from heavy investments in, say, nuclear power plants, coal strip mining, or off-shore wind parks, the respective economic, ecological, and political cost-benefit ratio informs future decisions. The formation of new political institutions under the impact of experimental research strategies is still in its initial stage. It seems that politics itself is changing as it acquires a more experimental style that—with respect to the European Union—is variously called "experimentalist governance," "regulatory experimentalism," or "collective experimentation" (Sabel and Zeitlin 2007; Yuval and Lezaun 2006; Felt and Wynne 2007). These moves to embrace new innovation strategies indicate that the revision of the Baconian contract toward a knowledge-society contract is under way.

Our vision of an epochal break advances a rather modest view with respect to fundamental cognitive shifts. It emphasizes the opportunities and politics of experimental practices. We are well aware of the conflict between the steadily growing number of more or less explicit projects of field experimentation on the one hand and the precarious disclaimer of social acceptance and institutional responsibility on the other. Thus the knowledge society finds itself in a paradoxical state: the more it absorbs scientific knowledge, the more it is compelled to deal with nonknowledge and its variants (ambiguity, ambivalence, indeterminacy). If experimentation is a privileged way out of this dilemma, the institutional tools for political use need to be further developed. Politics is bound to the rhetoric of right and wrong, and public opinion is strongly oriented toward preventing and avoiding risks. An experimental attitude, however, requires hypothetical reasoning and runs the risk of failure. Unlike the warnings heralded by the advocates of the risk society, where failures are seen mainly as malfunctions of the system, the knowledge society conceives failures as constructive components of learning.

Certainly there are cases where this kind of learning does not seem to be very clever or even appears cynical (society as a population of guinea pigs). Anthropogenic climate change would have been a case in point if it had been planned as a global experiment. Today juridical norms set limits to experimenting with uninformed people. Ethical norms should prevent the burden of risks being shifted onto future generations. However, in most innovative technological fields such as nano-, eco-, or biotechnologies, human-machine communication, medical research, energy transformation, and conservation, we are witnessing a merging of experimental application and basic research. This is the arena where an institutional and conceptual framework is needed to provide a proper conceptualization of field experimentation comprising epistemic norms, political guidelines, and policy procedures. Experimental governance would include making explicit the experimental design, establishing monitoring and data processing, keeping the public informed, and negotiating with concerned people and interest groups or, even better, enabling them to participate.

Whether these conditions slow down or accelerate innovation strategies is hard to predict. In Bacon's day it quickly became clear that the Baconian contract not only gives science its indispensable, independent, and uncontrolled space but also fulfills the expectations of the contractual parties with respect to the added value of "fruits of knowledge." In our own times the epochal break is defined, first by the (still precarious) acceptance of science as an agent of societal change that turns (parts of) science into field experimentation, and second by growing demands for public engagement to shape and control knowledge production to turn (parts of) science into a democratic endeavor. Obviously the new contract between science and society is being forged in an experimental mode.

NOTES

1. We are fully aware of the disciplinary use of the concept of "field experiment" in social anthropology or ecology, for example (Groß, Hoffmann-Riem, and Krohn 2005, 16). At the same time, our suggestion to use "field experiment" in a broader and rather more systematic sense is supported by other initiatives, such as the Network for the History and Sociology of Fieldwork and Scientific Expeditions (see http://www.fieldstudies.dk/).

2. Ian Hacking's *Representing and Intervening* (1983) is seen as the first and most influential book pointing to the proper life of experiments.

3. K. A. Appiah noted at his presidential address of the APA Eastern Division: "The recent return to these shores of the epithet 'experimental philosophy' is—as one tendency in our profession might put it—a return of the repressed."

4. See also Heidelberger and Steinle 1998.

5. This purported option is clearly rejected as a nonoption in the article by Jaqueline Cramer and Wolfgang van den Daele (1985). They point out that scientific ecology can plead just as little for avoiding the reductionist path as physics can. They propose instead a bifurcation of ecology: the scientific and the technological path. Only this latter one, the technological path, allows for what they call normative natural knowledge.

6. This is the title of an article by Wolfgang Krohn and Johannes Weyer, first published in German in 1989 (and in English in 1994).

7. From aquatic ecologist Walter Geller in his abstract for the ZiF workshop "From Lab to Field: Transforming Research Practices," Bielefeld, Germany, July 7, 2007.

8. See, for instance, Einsele 1936—one of the first articles in a series.

9. The institute was renamed MPI for Evolutionary Biology in March 2007.

10. They are not part of the native flora but come from abroad and spread in their new environment.

11. See "Population Biology of Alien Invasive Plants," online at http://www.uni bielefeld.de/biologie/Oekosystembiologie/doc/oeko25.html.

REFERENCES

Bacon, Francis. 1860 [1962]. "Novum Organum." In *The Works of Francis Bacon*, vol. 4, collected and translated by R. Ellis, J. Spedding, and D. D. Heath. Reprint of the London 1860 edition. Stuttgart-Bad Cannstatt: Frommann-Holzboog.

Böhme, Gernot. 1979. "Alternatives in Science—Alternatives to Science?" In *Counter-Movements in the Sciences: The Sociology of the Alternatives to Big Science*, edited by Helga Nowotny and Hilary Rose, 105–25. Dordrecht: D. Reidel.

Campbell, Donald. 1969. "Reforms as Experiments." *American Psychologist* 24: 409–29.

Carrier, Martin. 2006. *Wissenschaftstheorie: Zur Einführung*. Hamburg: Junius.

———. 2007. "Theories for Use: On the Bearing of Basic Science on Practical Problems." Paper presented at First Conference of the European Philosophy of Science Association. Madrid, May 15–17. Available online via the University of Pittsburgh's Phil-Sci Archive, http://philsci-archive.pitt.edu/3690/.

Cramer, Jacqueline, and Wolfgang van den Daele. 1985. "Is Ecology an 'Alternative' Science?" *Synthese* 65: 347–75.

Dewey, John. 1929. *Context and Thought*. Berkeley: University of California Publications in Philosophy.

Einsele, Werner. 1936. "Über die Bedeutung des Eisens, des pH und der Fäulnisvorgänge für den Kreislauf des Phosphats." *Fischereizeitung* 39: 1–8.

Felt, Ulrike, and Brian Wynne. 2007. *Taking European Knowledge Society Seriously*. Brussels: Directorate-General for Research Science, Economy and Society EUR 22700.

Gibbons, Michael, Camille Limoges, Helga Nowotny, Simon Schwartzmann, Peter Scott, and Martin Trow. 1994. *The New Production of Knowledge: The Dynamics of Science and Research in Contemporary Societies*. London: Sage.

Groß, Matthias, Holger Hoffmann-Riem, and Wolfgang Krohn. 2005. "Realexperimente: Ökologische Gestaltungsprozesse in der Wissensgesellschaft." Transcript. Bielefeld, Germany.

Hacking, Ian. 1983. *Representing and Intervening*. Cambridge: Cambridge University Press.

Heidelberger, Michael, and Friedrich Steinle, eds. 1998. *Experimental Essays—Versuche zum Experiment*. Baden-Baden: Nomos.

Knorr-Cetina, Karin. 1999. "Die Manufaktur der Natur—Oder: Die alterierten Naturen der Naturwissenschaft." IWT-Paper, 23. In *Die "Natur" der Natur*, edited by Institut für Wissenschafts- und Technikforschung, 104–19.

Kohler, Robert E. 2002. *Landscapes and Labscapes: Exploring the Lab-Field Frontier in Biology*. Chicago: University of Chicago Press.

Krohn, Wolfgang, and Johannes Weyer. 1994. "Society as a Laboratory: The Social Risks of Experimental Research." *Science and Public Policy* 21: 173–83. First published in German in 1989.

Krohn, Wolfgang, and Wolfgang van den Daele. 1998. "Science as an Agent of Change: Finalization and Experimental Implementation." *Social Science Information* 37: 191–222.

Sabel, Charles, and Johathan Zeitlin. 2007. "Learning from Difference: The New Architecture of Experimentalist Governance in the EU." *European Law Journal* 13: 271–327.

Schäfer, Lothar. 1999. *Das Bacon-Projekt: Von der Erkenntnis, Nutzung, und Schonung der Natur.* Frankfurt am Main: Suhrkamp.

Serres, Michel. 2000. *Retour au contrat naturel.* Paris: Bibliothèque Nationale de France.

Small, Albion. 1921. "The Future of Sociology." *Publications of the American Sociological Society* 16: 174–93.

Stokes, Donald. 1997. *Pasteur's Quadrant: Basic Science and Technological Innovation.* Washington, D.C.: Brookings Institution Press.

Yuval, Millo, and Javier Lezaun. 2006. "Regulatory Experiments: Genetically Modified Crops and Financial Derivatives on Trial." *Science and Public Policy* 33: 179–90.

Intensification, Not Transformation
Digital Media's Effects on Scientific Practice

VALERIE HANSON

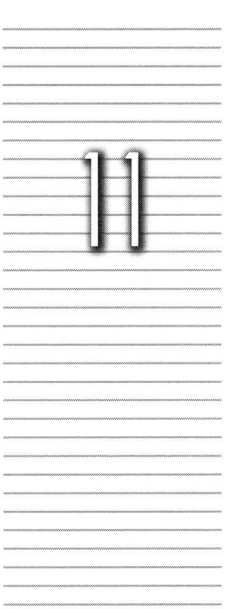

PART OF THE CURRENT CONTEXT in which scientific practice occurs is the increasingly frequent use of digital media to produce and communicate knowledge. Of course, this context is not science's alone; rather, it permeates many aspects of developed- and developing-world societies. Discussions about the significance of digital media's effects on how individuals access and interact with information through digital media—and further, how these interactions inform and form practices of knowing and communicating—have also become more frequent. In fact, some commentators claim that we are in the midst of the shift to a "digital age" or that we are undergoing a "digital revolution," even suggesting that digital media use has caused a transformative break from previous ways of making and communicating knowledge.[1] Although the extent of these claims about a societal shift from analog to digital are certainly debatable (and debated), the subject of these claims raises questions about how this context of the adoption and proliferation of digital media affects scientific practice. Has the adoption of recent new media such as digital images, websites, e-mail, and other electronic forms of communication changed science in relation to the practices of knowing and communicating that are integral to scientific practice? If so, how significant is this change?

Of course, when attempting to assess scientific change, the approach to take becomes a question. Alfred Nordmann (see his chapter in this edited volume) provides one useful way of articulating the issue of assessing change in terms of problems of distance. While Nordmann focuses on the shift from the scientific enterprise to a regime of technoscience, his comment about choosing a vantage point from which change may be visible that is neither too close nor too far is instructive here as well. For example, a close vantage point reveals the detailed historical and cultural contexts in which digital media have emerged and exist. These contexts are crucial to understanding specific changes in scientific practice; yet from this perspective, the larger question of change does not seem so important.

However, although choosing a distant vantage point that allows a wide perspective on the adoption of digital media may reveal general changes, it may overlook such details as disciplinary or cultural differences. Therefore, while the perspectives afforded by either a close or distant vantage point are important for analyzing and assessing the quality and scope of the work of digital media, neither provides a focus for determining the role of media in scientific change. This chapter attempts to navigate between these two positions by focusing on the medium in scientific practice. This approach can provide a bridge between the very specific and the very general by presenting insights into some of the practices of making and communicating knowledge in digital forms, practices that can then suggest possible ways to assess the significance of digital media for scientific communication and knowledge production.

"Media" is a slippery term with many definitions. Art history and English professor W. J. T. Mitchell provides a definition of media relevant to understanding its impact on scientific practices in his recent book *What Do Pictures Want?* Mitchell (2005, 198) builds on Raymond Williams's (1977, 158–64) definition of media, describing the term as referring to "not just materials but . . . material practices that involve technologies, skills, traditions and habits." This emphasis on media as a set of practices allows for analysis of media productions in a way that does not only focus on the product—an image, a text, or a film, for example—but also emphasizes the act of making in the scope of analysis, including the contexts of production as well as cultural elements of the product, such as conventions or tropes that both compose and are composed by the communications in which they appear. Such practices can then be analyzed for not only what the communication means, but also for how the medium and the technology that produced it influence the communication. (See Angela Krewani's chapter in this edited volume for another approach to including cultural contexts in the analysis of scientific media that primarily emphasizes technical aspects of the medium.) The emphasis on practices also includes other aspects related to communication practices, such as a communication's functions within other practices and discourses. Following such practices in the context of scientific knowledge making and communication can suggest ways in which these practices may influence or alter extant scientific practices. Therefore, a focus on practices can point to the significance of changes accompanying the use of digital media.

Although the adoption of new media has altered practices of interacting with information in some important ways, this change does not seem so drastic as to qualify for an "epochal break" or "revolution" in scientific practice on the whole (at least so far). These changes are not technically "novel" in that they do not introduce an entirely different way of practicing science; however, these changes do intensify certain extant practices of producing and communicating scientific knowledge. Such intensifications are significant in that they may cause qualitative—and possibly transformative—changes in some specific cases and may cause smaller, more incremental changes in other cases.[2] Of course, it is impossible to fully explain these points in this short chapter; instead I briefly illustrate some of the ways in which the changing context of application in relation to the shift to the digital affects scientific practice. I'll describe one main example of the use of digital images to highlight some of these characteristics, make a few points about the changing roles of other digital media as a partial context to the discussion about digital images, and then turn to the assessment of these changes' significance.

1. From Analog to Digital Practices: Informatic Images in Scientific Knowledge Production and Communication

Scientists have long presented information in visual forms, from drawings of natural observations to tables and charts, as part of their descriptions of experiments or observations used as evidence or documentation, or as illustrations of concepts. However, recently some visualizations that include scientific information have appeared to take on other functions in scientific practice as well. For example, the U.S. National Academy of Sciences Board on Chemical Sciences and Technology published a report on challenges and possibilities for chemical imaging.[3] The report lists among important challenges for imaging: "understand[ing] and control[ling] complex chemical structures and processes," "understanding and controlling self-assembly," and "understanding and controlling complex biological processes" (Board on Chemical Sciences and Technology 2006, 22–25). In this case imaging is used not only to record observations or collect data; it is also used to control what is imaged in the process of imaging. In other words, imaging can also allow researchers to interact with the object of study through data derived from the imaging process in a feedback loop back to the object, often in real time. This interaction with information through imaging can not only frame the object of study but also guide researchers' interactions with the object as they study it.

One of the reasons for expanded uses of imaging can be found in the development of visualization technologies that create images digitally, often by plotting data points derived through nonoptical measurement techniques such as ultrasound, CAT scanning, or even measuring the tunneling current or resistance of electrons on a surface (techniques used by the scanning tunneling microscope and atomic force microscope, respectively). These images are composed of a matrix of data points where each point presents a certain value that can be assessed individually or in relation to other points' values. Researchers can use these images to visualize what was previously invisible due to such factors as inaccessible locations (galaxies, or brain tumors in living people) or small size (atoms) and communicate these visions to others. Researchers can also use digital images to present information as an image, even if the information does not refer to visualizable phenomena. At the same time, many new visualization technologies include the added component of allowing researchers to use the digital image's presentation of data as an interface to interact with what is visualized, as the example above suggests. This development of digital media to use imaging for not only seeing and communicating but also interacting through a visualization with what was previously inaccessible or invisible raises questions about how such activities and the products of imaging may affect scientific practices.

The digital informatic image is not entirely new in some regards; computer-based digital informatic images share characteristics with scientific visualizations not made on a computer, such as charts, graphs, diagrams, tables, and maps. These shared characteristics make the scientific informatic image a useful case with which to study effects of the shift from analog to digital media. Indeed, a few characteristics of the digital production and viewing practices of these images illuminate some of the changed abilities and habits of working with the information presented in them. These characteristics include the increased amount and type of information that can be stored in an image, the intensity of the manipulation or arrangement in which the image creator engages, the extended time the image-maker spends determining the data's appearance to form an image, and the different possible interactions in which the image-viewer can engage.

Although the shift from analog to digital does not affect informatic images' overall function, to present information, digital images, like other forms of digital media, can present larger amounts of information in a digital space—in this case in the form of an image. For example, while a nondigital informatic image such as an anatomical drawing of a heart presents scientific information about what it shows, the image-maker presents visual data in terms of a representation of the object that is analog; that is, it presents data of "continuously variable, measurable, physical quantities, such as length, width, voltage, or pressure" as the image-maker perceives the data.[4] While a digital image of a heart can include a representation of the physical object, it can also present further information. The digital image is created through sampling the data so that each pixel contains a discrete numerical value that can be converted to another form of data, such as a graph, or a subsection can be analyzed in connection with its neighbors with greater precision.

These added capacities can change what the visualizations refer to—unlike such images as analog photographs or drawings, informatic digital images do not necessarily have a direct referent. Instead, as art historian Jonathan Crary (1990, 2) has remarked, "if these images can be said to refer to anything, it is to millions of bits of electronic mathematical data." This also allows the matrix composing the image to function as a database that researchers can search for particular values. Also, further complicating any connection to a referent, as Gilles Deleuze (1989, 265) has observed, each or a group of the pixels composing a digital image can form an image, which then allows for unlimited production of other images from the parts of the image. Digital informatic images, then, do not always present a view of an object; instead, they can also function more like tables or graphs in presenting a spatial arrangement of data. However, digital informatic images can also present a representation of an object and so at times resemble analog images such as photographs if the image creators so choose, which adds to the range of functions a digital informatic image can perform.

This expanded choice in arranging the data visually forms another difference between analog and digital informatic images. Although a creator of an analog image such as an anatomical drawing can of course choose how to arrange the data, she or he often does so by eye as she or he is composing the image or before starting to draw. In contrast, digital image-makers can take advantage of the fact that digital images are composed of discrete numbers, so they are able to arrange the information by eye or by computer algorithm, pixel by pixel. This lets digital image-makers manipulate the image's characteristics in expanded ways as they work with the full set of data. These options for arrangement and manipulation are also intensified because image creators can change parameters of how the data are expressed through computer algorithms, so researchers can apply changes quickly and consistently. For example, researchers can decide how to label data values with hue or intensity to highlight what they notice in the data and can quickly make many images with the same data by altering this labeling. Image-makers can thus determine the visual appearance of an image's data—what the image presents—through image processing, without changing the data itself. Therefore, uses of different visual labels can present entirely different images while using the same data matrix. This allows researchers to make more decisions about how to present their data; conversely, this means that researchers face a wider array of choices for presenting the data.

This ability to arrange the information and so manipulate the image also means that researchers participate in the imaging process in an exploratory, time-intensive way. While those creating a nondigital image, such as an anatomical drawing, also can arrange their data as they form the image, they do not have the same options in working with the full collection of presented data (the whole drawing, for example) to recompose it in the same intensified process. Michael Lynch's (1991, 72) explanation of the extended process in his study of astronomers' digital image productions provides a sense of what occurs, for example: "The real-time work of digital image processing involves a play at the keyboard, where images on the monitor are continuously recomposed by changing the palette, using touch-screen routines, plugging in parameters, and trying out different software manipulations."

This extended time while the image is being composed allows researchers to continue interacting through the image with the data and different imaging techniques, trying techniques and then responding to what they do. This increased intensity is due in part because the digital interface not only presents imaged data in a structure (the imaging software, for example) but also includes a structure for interaction through the computer software and hardware. Ann Johnson and Johannes Lenhard (see their chapter in this edited volume) articulate the role of the personal computer in encouraging an exploratory mode of research for those using simulations and modeling: the computer interface also creates an exploratory space in creating digital images because the data is presented fully but is alterable, which encourages further interaction with the data through the interface of the screen and imaging software.

Viewers of digital informatic images can also manipulate the data through image-processing programs if the image is on a computer screen. For example, viewers can either zoom in on a section or change the appearance to highlight different data. The image becomes an interface as viewers-become-users study its parts, and if on a computer, change them and produce a new image. In this way image-viewers become more intensively immersed in the image, similar to image-makers, and also have a different interaction with the information on the screen than in a nonmanipulable image. This may occur regardless of the viewers' knowledge level of the scientific data, because viewers are familiar with this characteristic of the medium due to the manifestation of digital images as interfaces in so many cultural and social contexts. (See Krewani's chapter in this edited volume for a more detailed analysis of the technical aspects of the medium regarding digital film, especially as viewed on a computer.) The emphasis on change among data points presented in digital images through color gradations or intensity invites viewers to move through the image and virtually go through the process of experiencing patterns of data. Because the image functions to present a matrix of data as opposed to a more conventional representation, such as an optical photograph, for example, viewers' purposes for looking at an image may also be different: viewers may look at the overall shape or object presented (if an object is presented) or they may look at some data in particular, or even compare certain bits of data, using the image as a database. Arrangement of data, as with other digital media, then becomes more flexible and alterable in digital images than in nondigital images for both image-creators and image-viewers as they use the image as an interface with the data presented.

These are a few ways that digital visualization technologies have expanded the functions of images and imaging so that researchers can use them to think through problems by interacting with data on the screen and in the image. These technologies also allow researchers to interact with experimental objects, such as the molecules and atoms to which the Board on Chemical Sciences and Technology (2006) referred through the data collected by visualization technologies. The experiment itself can occur through the imaging interface, so imaging becomes an experimental method. Furthermore, images not only demonstrate evidence of phenomena but also become an experimental space, an interface between researcher and object of study.

This brief exploration of scientific informatic digital images reveals not novel but intensified ways of arranging, presenting, and interacting with information that encourage image-creators and image-viewers to engage in qualitatively different habits of reading, using, and interacting with information and the medium of the digital image. These habits are not based on contemplation of a fixed arrangement of data, but on interaction; it is this experiential interaction with the data that helps provide a sense of the information's significance. In this way scientific digital informatic image-makers' and image-viewers' habits of interaction with the information presented in the image differ from those used in creating or viewing an analog image.

2. Assessing the Significance of the Digital Context of Application

Given this example of the digital informatic image, with consideration of the "medium" as a set of practices, what conclusions can be drawn about the significance of such intensified ways of arranging, presenting, and interacting with information? The focus on practices in which users of a medium and viewers of work in that medium engage allows us to navigate between individual productions (such as a particular image) and abstractions to a certain degree. Overall, these practices do not stand out as entirely novel or revolutionary: after all, scientists still plan and carry out experiments, and scientists still use other means of producing and communicating knowledge. In addition, there is some evidence that digital media use becomes incorporated within extant structures of knowledge and practice in specific scientific communities, as opposed to presenting novel practices (see Covi 2000 and Merz 2006, for example). And yet these intensifications of existing scientific practices do indicate changes in ways of interacting with and responding to information and, through this information, the framing and study of the objects of science. This conclusion is fairly general; however, it can generate some directions for further research on media practices in more specific contexts. The focus on practices can then provide a partial framework for further assessing the significance of scientific change.

First, the examination of practices associated with a medium encourages questions of whether the identified practices are isolated acts or are related to other practices in scientific knowledge production and communication. A quick glance at some examples of other digital media use reveals that the practices associated with the digital informatic image do not exist on their own. Instead, they are imbricated with other practices of knowledge making and communication that themselves are intensified and amplified in some similar ways. Other digital media practices also can intensify the amount of information available to users; like digital visual images, electronic versions of scientific papers and data can circulate relatively quickly and (often) cheaply over e-mail or in electronic versions of journals. The website ArXiv.org, hosted by Cornell University, for example, holds "620,005 e-prints in Physics, Mathematics, Computer Science, Quantitative Biology, Quantitative Finance and Statistics" as of August 10,

2010—all submitted by researchers in those fields at the time they submit articles to journals, and all freely accessible over the Internet.[5]

This changed context of knowledge circulation also means that researchers may have to interact with information in different ways, such as running searches through archives for specific topics to learn about advances in their fields. Like digital informatic images, digital media such as electronic papers and e-mail can lead to intensified manipulation of data, as researchers move through greater amounts of information to find what they need. They may also spend more time searching databases and articles to find information. The availability of scientific knowledge online and in electronic database form has also affected the shape of science, as it has helped foster new fields or subfields, such as biomathematics or theoretical biology in which scientists use vast data sets such as genomes to conduct research.

Electronic versions of articles and databases such as ArXiv.org can also significantly increase access to scientific knowledge, which then affects scientific practices in a few ways. Scientists worldwide can access the same information, which fosters global growth in fields, as opposed to limiting access to research to specific institutions. In addition to allowing for wider participation in scientific fields, electronic communication pathways increase the ease of interaction between scientists, who can communicate more information more quickly with others around the world and thus collaborate based on interest or specialization regardless of where the scientists live. In addition, these communication pathways are not reserved for science—more is accessible to nonscientists online, for example. Therefore, information can travel beyond disciplinary boundaries more easily, so that, for instance, stakeholders such as patients or activists can access scientific studies and original research. Given this focus on intensified practices, the attention not to what is wholly novel but to what is different encourages questions about whether the identified practices have existed all along but have not been visible for some reason. For example, what sorts of interactions with data already existed that did not seem so significant until the adoption of the digital informatic image?

The focus on media practices also lends itself to asking questions about the effects of these practices. For example, in the case of digital informatic images, what does it mean for scientific practice that digital image-makers and image-viewers are deploying a visual language to describe and to communicate about experiments, and to interact with data—and even objects of study? What effects on science does developing this more visual, experiential knowledge of data for viewers have so that direction through the data is signified by changes in the image, for example, as opposed to verbal or textual explanation? Also, what effects does this have on the data itself, in terms of framing of information? In terms of what data is collected, and what is not? How does this different frame for data affect the direction of experiments, lines of inquiry, and possibly of fields? What practices do those used in the creation and understanding of digital media replace?

In addition, one can assess significance by examining how the uses of digital media are rhetorical: what is persuasive about the uses of digital media in both producing and communicating knowledge? Differences in communicative strategies suggest that as certain aspects of a new medium or communication technology allow for further functions or abilities, what becomes persuasive about communication in that medium may differ from communication in more conventional media. Indeed, what becomes persuasive may differ from field to field as well. When does digital media become a more persuasive choice for researchers, and why? When does a digital image, for example, prove more persuasive to its viewers than an analog one? What elements or rhetorical conventions appear in digital media productions in a field, and how do they compare with that field's established practices of communication? Such lines of inquiry can reveal shifts in communication patterns, and also in what is communicated, in more specific studies that take historical and social contexts into account as part of their examination.

Assessing significance also encourages asking what the benefits are for asserting that such changes are significant in some way. A focus on practices shifts emphasis from the constant search for the "novel" that can distract from the more subtle changes that may be occurring, while at the same time still allowing for a focus on changes in science in the context of application through understanding the nature of these changes—in this case, these intensifications. As gradual or incremental or non-"revolutionary" changes often slip under the radar, science may not realize that some of the context of application is shifting—or that it has already changed in ways that may have profound, yet not fully transformative, effects on scientific practice.

NOTES

1. Such claims are widespread in different fields, as a simple Internet search will show. For example, many media and popular studies commentators follow Marshall McLuhan's (1962 and 1964) emphasis on the social and cultural importance of media to proclaim a revolution or new digital age: see Manovich 2001, 19, and Johnson 1997, 40, for example. Some studying the rhetorical aspects of the digital

also take for granted that such a "revolution" has occurred: see Johnson-Eilola 2005 and Rice 2007 for a few recent examples.

2. My focus on intensities of practice draws from rhetorical studies' focus on the effects of communications (or elements therein) as well as from philosophical works that foreground intensities and flows of forces (see Deleuze and Guattari 1987, for example). Following Friedrich Nietzsche, Michel Foucault also emphasizes intensities and forces in his work (see Foucault 1976 and 1977, for example; also see Nealon 2008 for a discussion of intensification of power in Foucault).

3. The report defines chemical imaging as "the spatial (and temporal) identification and characterization of the molecular chemical composition, structure, and dynamics of any given sample" (Board on Chemical Sciences and Technology 2006, 14).

4. This definition of "analog" is according to the 2004 online edition of *The American Heritage Dictionary of the English Language,* fourth edition, Houghton Mifflin Company; see http://dictionary.reference.com/browse/analog.

5. The ArXiv.org e-Print Archive is hosted by Cornell University; see http://arxiv.org/.

REFERENCES

Board on Chemical Sciences and Technology. 2006. *Visualizing Chemistry: The Progress and Promise of Advanced Chemical Imaging.* Washington, D.C.: National Academies Press.

Covi, Lisa M. 2000. "Debunking the Myth of the Nintendo Generation: How Doctoral Students Introduce New Electronic Communication Practices into University Research." *Journal of the American Society for Information Science* 51 (14): 1284–94.

Crary, Jonathan. 1990. *Techniques of the Observer: On Vision and Modernity in the Nineteenth Century.* Cambridge: MIT Press.

Deleuze, Gilles. 1989. *Cinema 2: The Time-Image,* translated by Hugh Tomlinson and Robert Galeta. Minneapolis: University of Minnesota Press.

Deleuze, Gilles, and Felix Guattari. 1987. *A Thousand Plateaus,* translated by Brian Massumi. Minneapolis: University of Minnesota Press.

Foucault, Michel. 1976 [1980]. *History of Sexuality.* Vol. 1., translated by Robert Hurley. New York: Vintage.

———. 1977. *Discipline and Punish: The Birth of the Prison,* translated by Alan Sheridan. New York: Vintage.

Johnson, Steven R. 1997. *Interface Culture.* New York: Harper.

Johnson-Eilola, Johndan. 2005. *Datacloud: Toward a New Theory of Online Work.* Cresskill, N.J.: Hampton Press.

Lynch, Michael. 1991. "Laboratory Space and the Technological Complex: An Investigation of Topical Contextures." *Science in Context* 4 (1): 51–78.

Manovich, Lev. 2001. *The Language of New Media.* Cambridge: MIT Press.

McLuhan, Marshall. 1962. *The Gutenberg Galaxy: The Making of Typographic Man.* Toronto: University of Toronto Press.

———. 1964. *Understanding Media: The Extensions of Man.* New York: McGraw Hill.

Merz, Martina. 2006. "Embedding Digital Infrastructure in Epistemic Culture." In *New Infrastructures for Knowledge Production: Understanding E-science*, edited by Christine Hine, 99–119. London: Information Science Publishing.

Mitchell, William J. T. 2005. *What Do Pictures Want? The Lives and Loves of Images.* Chicago: University of Chicago Press.

Nealon, Jeffrey T. 2008. *Foucault beyond Foucault: Power and Its Intensifications since 1984.* Stanford: Stanford University Press.

Rice, Jeff. 2007. *The Rhetoric of Cool: Composition Studies and New Media.* Carbondale: Southern Illinois University Press.

Williams, Raymond. 1977. "From Medium to Social Practice." In *Marxism and Literature*, 158–64. New York: Oxford University Press.

Technologies of Viewing
Aspects of Imaging in Natural Sciences

ANGELA KREWANI

"MEDIUM" IS A VERY BROAD TERM, denoting the transmission of a certain message with the help of specific tools. Its definition extends from technological, audiovisual media to speech, drawing, language, or writing. Accordingly, the history of science is also a history of the development of use of different media, especially in regard to the visualization of its objects. Traditionally, all natural sciences have used drawings and charts to specify their chosen examples. It is readily apparent that the connection between an object and its visual representation is not naturally given but is constructed by way of conventions. Contemporary science studies agree on the constructedness of the scientific image and regard the image as an effect of the interplay of varying factors, such as cultural and technical values. But representational media do not just construct the relation between images and objects, they also propose a certain way of perception and thus construct a relation between the image and perceiving subjects.

With its special techniques of editing and montage, film governs the perception of content and requires knowledge of the characteristic visual logic of the medium for deciphering the flow of images. Different media thus structure the perception of reality. Within media theory this structuring function is called a "dispositif"—the most famous example of this being the cinema as a means of physically configuring the observer and the image and of thus disposing the viewer to be overwhelmed by the powerful illusion of movement on the movie screen. This chapter develops this notion by considering the perceptive powers of media. In contrast to Valerie Hanson's chapter, which underlines pragmatic aspects of media use, this chapter focuses on the technical aspects of certain media in contemporary research practice and on the perceptual consequences of these technical features—focusing on the production, perception, and distribution of images and illustrations.

The long history of scientific illustrations is marked by two landslide events. First came the emergence of photography and film as technical media with the capacity to mimetically record an object. Second, the introduction of digital image rendering set up a new relationship toward the depicted object in that it does not rely on the material presence of the depicted object. Digital imaging is capable of transforming any set of data into a visual image. William Mitchell (1992, 163) labels it as "already an interpretation, a mise-en-image," which is a combination of digital technology, cultural projection, and representation. In her chapter Valerie Hanson scrutinizes this constellation with reference to new cultural formations of media. With regard to the claim of an epochal break in recent history of science and technoscience, the question emerges, how the transition from analog recording to digital imaging is affecting the dynamics of specific sciences and their representational strategies.

After discussing film and photography with a view to their visual and seemingly documentary qualities, basic tenets of film theory will be applied to the transition from the recorded (film) to the constructed (digital) image and the concomitant change in the practical employment of these media in science. Film theory and particularly the notion of technical and cultural "dispositifs" offers a flexible model to capture the emergent dynamics of media technologies in scientific and public culture.

From their beginnings, technical media such as film and photography have drawn the interest of scientists. Robert Koch, a leading microbiologist of his time, worked intensely with photography (Hennig 2004), and especially in medicine and biology film stepped early on into the center of the scientific visual imagination (Fox Keller 2002). Lisa Cartwright (1995, 1) reminds us of the fact that obituaries of Charles Lumière point to the importance of his invention for science. Aside from doing the serious work in the laboratory, scientific film has served as an entertaining spectacle in the cinemas. Scholars of early film have shown that there was no significant distinction in the presentation of fiction film and films dealing with scientific facts (Gunning 1990; Tsivian

1996; and Landecker 2006). Scientific film served both in the lab and as a fascinating spectacle, especially where it provided "insight" into the body, as the x-ray films demonstrate. With film for the first time in the history of scientific illustration, a medium was capable of recording movement, which was not possible by means of photography.

Scott Curtis (2004) expands on the astonishing capability of film to depict a body in motion. He notes the astounding analogy between medicine and film. In his view medicine moves between life and death in just the way in which film and photography travel between the static and the moving image. Curtis thereby offers a very special view on the use of film in its early stages. His starting point is the film's capacity to record movement that establishes a structural analogy between science and film: both are capable of isolating the single moment and of representing change in a succession of discrete elements. Thus, according to the philosopher Henri Bergson, science and film organize time. In Bergson's view contemporary science operates in analogy to the cinematographic mode by substituting the object with a sign (Curtis 2005). This brings about an unstable relationship between scientific object and its filmic representation.

Although Bergson's remarks are very general and he specifies neither the scientific discipline nor the media technology, he nevertheless points to a highly topical development in a variety of contemporary scientific disciplines— namely, the increasing quantity of visual data in the production of scientific knowledge. Especially technologies such as microcinematography and x-ray cinematography have offered a detailed insight into the human body. Slowly these visuals, which were simply representations of the inner human body and followed strictly codified rules of organized visual knowledge, became a focus of scientific reflection and spawned a confusion of material object and visual representation. Micrography, cinematography, and digital imaging have brought about forms of highly efficient representation of scientific objects. Especially in the realm of microscopic depiction and the transformation of numerical data into visual images, digital image rendering has become highly proficient. Some disciplines, such as nanotechnology, mostly because of the small size of the research object, rely on its visual representation. In this field a confusion of representation and material object is very likely to happen and introduces new and complex epistemological issues.

Against this epistemological backdrop of the move from recording to imaging, traditional cinematographic technology resolves the representational status of the scientific image: the filmic image does not document a scientific procedure, but it is part of the procedure itself, for example, in that it creates the discernible discreteness of moments in a continuity of motion.[1] Curtis exemplifies this fact with reference to Brownian motion, which was used for experiments on a molecular level. Molecules could not be represented photographically, since their movement was too fast. The experimenter set out to film a specific movement of the molecule, stopped the camera, filmed another movement, and so on. This technique or, in filmic terms, simulation of movement produced images of a process that could not have been documented before (Curtis 2004).

A similar process has been set up in microcinematography, which is the filming of material seen through the microscope. Hannah Landecker understands microcinematography as an early media technology to reach and document phenomena "below the threshold of perception." In her view film was so closely bound to scientific discovery that the camera became a central scientific tool, carrying the same importance as the microscope itself (Landecker 2006, 123). This allows for the reciprocal conclusion as well. Just as film has been understood as a scientific tool, the microscope can be conceived as a visual medium. In the context of science the medium film came forth as technology that could produce and record movement of all kinds, which could not have been documented before. It is therefore quite obvious that the media technology is responsible for the image and at least to some extent for the object of study.

Up to now, studies of the meaning of film in scientific contexts have concerned themselves either with the impact of film on science or with the structural analogies between film and scientific discipline. In other words, film does not only offer moving images, it also structures the viewing process and the viewer's perception. Relying on this understanding of the structural powers of the medium, I intend to open up another line of argument, which has not been considered so far in the study of media in science. As mentioned already, technological media have the capacity of structuring the viewer's perception of the disclosed event. This capacity has been exhaustively scrutinized within film theory under the concept "dispositif" informing us about the perceptive connection between the viewer and the viewed.

Over the past years film theory has developed various concepts of the visual dispositif, stretching from the exclusively technical to the social implications of the term. The strong technological impact is unraveled by Jean Louis Baudry (1986) and delineates the technological conditions of the filmic apparatus, which consists of the screen, the viewing subject, and the projector shedding light onto the screen. With the help of psychoanalysis, Baudry claims that cinematic technology is the visual basis for the construction of the viewing subject. Parallel to psychoanalytic positions of the emergence of human identity, Baudry supposes the identification of the viewer with the shadows on the silver screen. As a result of its psychic structure, the human subject has to identify with the characters on the silver screen, taking them as mirror images of him- or herself.

The pivotal significance of Baudry's concept of the dispositif is the transparency of the apparatus. Although the viewer is caught up in a technological setting, the setting as well as the editing of film aim at being transparent: the viewer is supposed to forget the technological precondition and to become absorbed in the filmic narrative. More so than the similarity in the treatment of time, is the transparency of the apparatus the common denominator of filmic image and science image (van Dijck 2005, 3–17). In the same way as the filmic apparatus intends to render its technology invisible, early medical imaging as well as image processing in the most advanced natural sciences render the technical apparatus invisible and focus on the image as "natural" object of science.

The second, more culturally oriented understanding of dispositif derives from a critique of the ideology of film. Jean Louis Comolli claims that general cultural contexts precede filmic technology. He thereby highlights the notion that film is always governed by the social structure of its society (Comolli 1980).[2] In Comolli's sense, the notion of the dispositif expresses the heterogeneity of technical and social techniques, their interactions, and their mutual influences. Although he developed this notion in film theory, it should be transferred to scientific disciplines working with visual apparatuses. Comolli stresses technology's capacity to inscribe itself into the visual image. Contrary to Baudry, Comolli highlights the cultural aspects of a certain technology.[3]

Cultural and social values shape the respective technology and define its use. Comolli refers to Hollywood Cinema, which in his account functions as an ideological machine. Widening this understanding of the social as constructing meaning and structuring the use of technology, Comolli's understanding of the term "dispositif" comes very close to Michel Foucault's understanding of it. Gilles Deleuze picks up on the concept of dispositif as formulated by Foucault and points to the general processual energy of the term, thus highlighting its dynamic aspects: On the one hand, it safeguards established forms of knowledge, subject-constitution, and social organization; on the other hand, it allows for new dynamics and partially accounts for its self-organization. In the sense of Deleuze (1991, 154), the dispositif is a flexible structure, organizing a set of different structures along traditional lines and undermining them simultaneously.

Summarizing these increasingly comprehensive appeals to the dispositif, we come to understand that we are dealing with a structure that helps us understand the impact of technology and society on the construction of the scientific image. Taking this structure as the analytical tool, it accomplishes an explanation of the complex and heterogeneous relationship between visual representation, technologies of the visual and society. I therefore propose an approach to the visual image in science that is informed by the dispositif, since this allows for the precise description of the technical tools and their visual implications. This approach becomes especially important when we are facing the multifarious relations of digital visual imagery in contemporary sciences such as nanotechnology or biotechnology, where a complex process of digital imaging steps in between the researcher and the image.

Interestingly enough, though the concept of the dispositif is not used much within science studies, it is often indirectly hinted at. Timothy Lenoir comes quite close to the idea that society and technology somehow shape the scientific process and its visual representation. But his approach, being based on Derrida's idea of writing and dissemination, tackles the concept of representation as mimesis of an outside object (Lenoir 1998). Although referring to postmodern philosophy, Lenoir is lacking a precise approach to the image, which captures the semantic and technological implications of scientific media. By adopting the metaphorical concept of inscription, he falls short of offering a literal understanding of the coming together of representation and technology.

Along the same lines, Hans-Jörg Rheinberger concentrates on the static and moving image. According to Rheinberger (1997a), the image's meaning does not come into being through an inherent connection with the represented, but it is shaped by the technological conditions and the cultural context. Lorraine Daston and Peter Galison came very close to the notion of the machinations of the dispositif, when they remark on the epistemological paradigms of radiology, which attempt to "eliminate the mediating presence of the observer," which in this case is the seeming elimination of the observing and recording technology (Daston and Galison 1992, in Pasveer 2006). Where all these accounts fall short is in the recognition that a representational technology like cinematography can simultaneously produce and record something by documenting the effects it creates—and that these effects are not therefore mere artifacts or constructs but owe to the intimate engagement of the technology with reality.

Against this background the cultural and technological understanding of dispositif comes very close to a conception of scientific knowledge as being brought about by the social and technological processes governing scientific work. The dispositif can be made visible through an analysis of the respective structures. What Rheinberger, Michael Lynch (1994), Daston and Galison (1992), and Bernike Pasveer (2006) label as the structure of scientific practice with the interaction of different disciplines, cultures, technologies, and social groups comes very close to the notion of dispositif. Considering the dispositif as an underlying organizational structure of scientific knowledge, the specific technology of image processing assumes utmost importance. Thus scientific imagery possesses a different epistemic status with respect to different imaging technologies, like drawing, photography, film, digital imaging, or hybrid forms of image processing.

Next I focus on the scientific image that is brought about by film in contrast to digital imaging. My central interest is to describe a change of the dispositif in the transformation from filmic recording to digital processing. The filmic image possesses a variety of characteristics that are transformed in the transition to the digital image. Contrary to manual illustration and photography, film was capable of documenting a real object—in this respect like photography—and of shaping something in the process of recording, primarily through the technique of montage. Film thus relies, like photography, on the actual "outside" object being represented and an "inner" media technology that is shaping the respective image. In this regard, film is in between mere documentation and deliberate rendering of an image. Montage technology is the first step toward a technological shaping of visual processes—while photography still relies on a material object even when the social aspects are taken into account (Rosenblum 1978).

With regard to the epistemic status of the filmic image, it can be stated that it records an outside object and that it can also be described as a hybrid between documentary photography and inventive digital rendering. As we have seen, scientific film has especially relied on the montage technology of film. But, and this tends to be overlooked by Scott Curtis (2004) and others, the reception of the filmic situation is governed by the technological apparatus. An approach that does not include the cinematographic apparatus of film falls short of explaining the representational power of the medium. The technological structure delineated by Baudry (1986) organizes the representational effect of the medium and serves to explain the convincing power of the montage technique even in presenting the apparent visual and phenomenal immediacy of observed processes. A pivotal aspect in the traditional use of film within scientific contexts is the established distance between viewer and the screen. The distance explains the perceptual powers of the filmic illusion. We thus are dealing with a highly organized and technologically constructed way of perception (Sobchack 1988).

The introduction of digital imaging produced changes in the technological and social dispositif of scientific imaging that force us to consider the following aspects. In the context of the new technology, imaging moved away from the filmic apparatus with its silver screen to another screen, which strongly resembles a television screen. Meanwhile, the continuing emergence of television introduced a new dispositif to the media world. The medium of television has created a different dispositif since the light for the image is not frontally projected but emerges from within the screen. Accordingly, the subject is differently positioned within the viewing process, and usually it is not fixed in front of the screen in a dark room. Instead, there is additional light in the room that comes from the screen, and the viewer can move about, which is only the most rudimentary of interacting with the digital image.

Since the emergence of television, the media situation has been characterized by an increasing hybridization of forms and technologies. The introduction of the personal computer added to this hybridization of media and forms, since the personal computer unifies most of the traditional media functions but also exhibits a different dispositif: the light is coming from the screen, usually there is a keyboard and, most important in this context, there is no longer a set distance between the viewer and the visual apparatus.

A similar movement or transition can be observed in respect to science: film (as analog medium) has left the focus of scientific attention and has been replaced by the screen, be it the computer screen or the screen of the microscope.[4] Crucial with this transfer from the silver screen to the computer screen is the change in the technological dispositif, which also effects a change of cognitive structure. Whereas with film, the technical components of scientific knowledge had been separated from the phenomena and from the viewer, they are conflated by contemporary technology. This means that the research and the experience of images occur in a different situation that strongly reflects the changed technological preconditions.

Accordingly, the separation of viewer, viewed, and viewing technology is hindered further by the use of the same instrument. The computer has to serve as an instrument of calculation and interpretation and as an instrument of observation and visual display at the same time. This blurring of technical conditions is an important factor in Nicolas Rasmussen's (1997) work on the electron microscope. In some ways Rasmussen is indirectly referring to the idea of a technical dispositif within the microscope. He conceives the lack of distance between visual screen and observing subject as a changed impact on the scientist's body and proposes a changed scale in human experience. The electron microscope alters the experienced scale of the observer's body. He or she is made tiny by proximity with small things (when imagining standing next to a treelike hair shaft), and by the same token made huge (when stretching from New York to Minnesota). Simultaneously, the microscope adds a change of place to this transformation of scale, much as Galileo's telescope brought the moon out of the heavens and down to the earth (Rasmussen 1997, 225).

Rasmussen goes on and describes a blending of microscope and researcher, who develops an increasing familiarity with the instruments and makes it personal. "The instrument became decreasingly refractive to their wishes, and increasingly predictable in performance" (Rasmussen 1997, 229). Rasmussen's observations on the conflation of technology and the human body come very close to Marshall McLuhan's (1964) understanding of media as extensions of the human body. In a similar manner but with a different terminology, Rheinberger relocates the creation of images to the inside of the microscope. Following Rheinberger (1997b, 274), the image is constructed with the help of image-makers and the microscope.

Outside of media terminology, this blending of instrument, researcher, and reality can also be termed as a "collapse of distance." Alfred Nordmann employs this term to refer to the singular situation of technosciences. For Nordmann the collapse of distance describes the lack of difference between reality and representation—and similar to the argument presented earlier in this chapter, Nordmann (2006) underlines this by pointing to the fusion of researcher and visual technology. In his introductory chapter in this edited volume, Nordmann picks up on this idea and claims the "impossibility to separate out the theoretical representation of nature and the technical intervention into nature." As far as the visual representations of nature are concerned, this impossibility of separation between "nature" and "technology" is the effect of an ongoing process in science, which reaches a new peak with the invention of technological and digital image construction (Daston and Galison 1992).

All this suggests that the image has assumed a new set of characteristics or, in terms of science studies, new epistemological values. The collapse of distance, the new technological dispositif, and the negation of the distance between the screen and the media subject creates a new epistemic position of the image as compared with photography and film (Rheinberger 1997a, 390). Assessing the new epistemological status of the image for scientific knowledge production, Lynch (1994) denies the representational function of the image and Timothy Lenoir (1998, 8) also requires a new evaluation of visual material: "Images and signals may be captured from cameras or sensors, transformed by image processing, and presented pictorially on hard or soft copy output. Abstractions of these visual representations can be transformed by computer visions to create symbolic representations in the form of symbols and structures. Using computer graphics, symbols or structure can be synthesized into visual representations." As a consequence, Lenoir (1998) envisions a completely new way of theorizing the scientific image, focusing on the possibility of its endless multiplication within a diverse technological context. Thus the status of knowledge is closely linked to the computer and its visual competencies. Research and image production have apparently entered a process of continual merging. Something like this seems to be implied in Rasmussen's remarks about the "personal dwelling" of the researcher.

It thus becomes evident that imaging in natural sciences has experienced a transformation with the introduction of technical media. To set up an analytical perspective on the impact of technologies of visualization, the concept of dispositif and the established relationship between visual apparatus and the viewer proves to be fruitful. Along with the technological change from filmic image to digital imaging, a change in "dispositif" has appeared. As was shown here, digital image processing has done away with the implied distance between image and viewer and instead brought about a merging of viewer and technology. This merging underlines the shift from scientific enterprise to technoscientific regime through an ongoing conflation of viewer, apparatus, and image.

NOTES

1 I understand this as a conceptual forerunner of microbiology the way Rheinberger conceives of it (Rheinberger 1997a).
2. "The tools always presuppose a machine, and the machine is always social before it is technical" (Comolli 1980, 122).
3 Comolli's definition of a "dispositif" comes very close to Raymond Williams's understanding of media practice as cited in Valerie Hanson's chapter in this edited volume.
4. Film has remained a very important medium in the communication of scientific results to a wider public audience. See Kirby 2003.

REFERENCES

Baudry, Jean Louis. 1986. "The Apparatus: Metapsychological Approaches to the Impression of Reality in Cinema." In *Narrative, Apparatus, Ideology: A Film Theory Reader,* edited by P. Rosen, 299–318. New York: Columbia University Press.
Borck, Cornelius. 2001. "Die Unhintergehbarkeit des Bildschirms." In *Mit dem Auge denken: Strategien der Sichtbarmachung in wissenschaftlichen und virtuellen Welten,* edited by J. Huber and B. Heintz, 383–86. Zürich: Edition Voldemeer.
Cartwright, Lisa. 1995. *Screening the Body: Tracing Medicine's Visual Culture.* Minneapolis: University of Minneapolis Press.
Comolli, Jean Louis. 1980. "Machines of the Visible." In *The Cinematic Apparatus,* edited by S. Heath and Teresa de Lauretis, 121–42. New York: St. Martin's Press.

Curtis, Scott. 2004. "Still/Moving: Digital Imaging and Medical Hermeneutics." In *Memory Bytes: History, Technology, and Digital Culture*, edited by A. Geil and L. Rabinovitz, 218–54. Durham, N.C.: Duke University Press.

———. 2005. "Die kinematographische Methode: Das 'Bewegte Bild' und die Brownsche Bewegung." *Montage a/v* 14 (2): 22–43.

Daston, Lorraine, and Peter Galison. 1992. "The Image of Objectivity." *Representations* (special issue "Seeing Science") 40 (Fall): 81–128.

Deleuze, Gilles. 1991. "Was ist ein Dispositiv?" In *Spiele der Wahrheit: Michel Foucaults Denken*, edited by B. Waldenfels and F. Ewald, 153–62. Frankfurt am Main: Suhrkamp.

Dijck, José van. 2005. *The Transparent Body: A Cultural Analysis of Medical Imaging*. Seattle: University of Washington Press.

Fox Keller, Evelyn. 2002. *Making Sense of Life: Explaining Biological Development with Models, Metaphors, and Machines*. Cambridge: Harvard University Press.

Gunning, Tom. 1990. *The Cinema of Attractions: Early Film, the Spectator and the Avant-Garde*. London: BFI Publishing.

Hennig, Jochen. 2004. "Vom Experiment zur Utopie: Bilder in der Nanotechnologie: Bildwelten des Wissens." *Kunsthistorisches Jahrbuch für Bildkritik* 2 (2): 9–18.

Kirby, David A. 2003. "Extrapolating Race in GATTACA: Genetic Passing, Identity, and the Science of Race." *Literature and Medicine* 23 (1): 184–200.

Landecker, Hannah. 2006. "Microcinematography and the History of Science and Film." *Isis* 97: 121–32.

Lenoir, Timothy. 1998. "Inscription Practices and Materialities of Communication." In *Inscribing Science: Scientific Texts and the Materiality of Communication*, edited by H. U. Gumbrecht and Timothy Lenoir, 1–10. Stanford: Stanford University Press.

Lynch, Michael. 1994. "Representation Is Overrated." *Configurations* 2 (1): 132–49.

McLuhan, Marshall. 1964. *Understanding Media: The Extensions of Man*. London: McGraw-Hill.

Mitchell, William J. 1992. *The Reconfigured Eye: Visual Truth in the Photographic Era*. Cambridge: MIT Press.

Nordmann, Alfred. 2006. "Collapse of Distance: Epistemic Strategies of Science and Technoscience." Lecture at the annual meeting of the Danish Philosophical Association, March. Copenhagen.

Panofsky, Ernst. 1974. "Die Perspektive als symbolische Form (1924)." In E. Panofsky, *Aufsätze zu Grundfragen der Kunstwissenschaft*, edited by H. Oberer and E. Verheyen. Second edition. Berlin: Bruno Hessling.

Pasveer, Bernike. 2006. "Representing or Mediating: A History and Philosophy of X-Ray Imaging in Medicine." In *Visual Cultures of Science*, edited by L. Pauwels, 41–62. Hanover, N.H.: University of Dartmouth Press.

Rasmussen, Nicolas. 1997. *Picture Control: The Electron Microscope and the Transformation of Biology in America 1940–1960*. Stanford: Stanford University Press.

Rheinberger, Hans-Jörg. 1997a. *Toward a History of Epistemic Things: Synthesizing Proteins in the Test Tube*. Stanford: Stanford University Press.

———. 1997b. "Von der Zelle zum Gen." In *Räume des Wissens: Repräsentationen, Codierung, Spur*, edited by M. Hagner and Hans-Jörg Rheinberger. Berlin: Akademie Verlag.

————. 2001. "Objekt und Repräsentation." In *Mit dem Auge denken: Strategien der Sichtbarmachung in wissenschaftlichen und virtuellen Welten*, edited by Jörg Huber and Bettina Heintz. New York: Springer.

Rosenblum, Barbara. 1978. *Photographers at Work: A Sociology of Photographic Styles*. New York: Holmes.

Sobchack, Vivian. 1988. "The Scene of the Screen: Beitrag zu einer Phänomenologie der 'Gegenwärtigkeit' im Film und in den Elektronischen Medien." In *Materialität der Kommunikation*, edited by Karl Ludwig Pfeiffer and Hans Ulrich Gumbrecht, 416–27. Frankfurt am Main: Suhrkamp.

Tsivian, Yuri. 1996. "Media Fantasies and Penetrating Vision." In *Laboratory of Dreams: The Russian Avant-Garde and Cultural Experiment*, edited by John Bowlt and Olga Matich. Stanford: Stanford University Press.

Technoscience as Popular Culture
On Pleasure, Consumer Technologies, and the Economy of Attention

JUTTA WEBER

The public consumes science for pleasure.

—MIKE MICHAEL

Far from depleting scientific materiality, worldliness, and authority in establishing knowledge, the "cultural" claim is about the presence, reality, dynamism, contingency, and thickness of technoscience.

—DONNA HARAWAY

THE INCREASING MARKET ORIENTATION of universities and other research institutions, the worldwide competition for key technologies, as well as the race for research funding and public attention are changing not only the relation between mass media and technosciences but also research strategies and paradigms of the technosciences themselves. In this chapter I analyze the cultural turn of technoscience(s) and changes in its epistemology, ontology, and rhetoric with regard to recent developments in personal service robotics and especially humanoid robotics. Robotics as a technoscience is not only more and more involved in PR activities, but is increasingly becoming—and stages itself as—an integral part of popular culture. This shift in robotics is accompanied by a move from the traditional expert culture of industrial robotics toward service robotics with its focus on consumer technologies and lifestyle products for everyday user(s).

Personal service robotics with human-robot interaction (HRI) focuses on affect, emotional bonding, and sociality of humans and machines, which is part of the cultural turn in robotics. Understanding "technoscience as an *integral* part of contemporary western culture" (Reinel 1999, 166) and taking its radical epistemological and ontological changes seriously, my main focus here

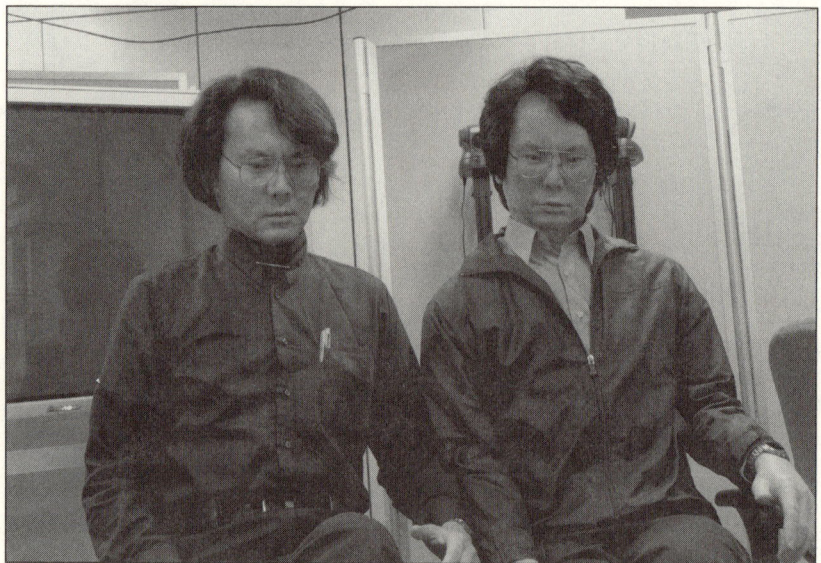

Figure 13.1. Roboticist Hiroshi Ishiguro with his "twin," the robot Geminoid, 2007. *Source:* Prof. Hiroshi Ishiguro and the ATR Intelligent Robotics and Communication Laboratories.

is less to discuss *whether* we live in the age of technoscience than to elaborate the transformation of robotics as a modern engineering field into a postmodern, "sexy," and media-effective technoscience rooted in pop and consumer culture (figure 13.1).

Toward Technoscience: Science and Technology as Cultural Practice

In the 1980s the cultural studies of science and technology—or cultural studies of technoscience—interpreted science and technology as "cultural practice and practical culture" (Haraway 1997, 66). In the face of the increasing hybridization of science, technology, industry, and society, this approach analyzed quite early the production of new concepts, meanings, and media in and through such technosciences as genetics, cybernetics, computer science, artificial intelligence, and reproductive medicine. One of cultural studies' central preoccupations was to show how the discourses and practices of the technosciences reshaped not only our political culture, kinship structures, and economy, but every dimension of everyday life (Reinel 1999).

Cultural studies of technoscience do not primarily interpret science as elite knowledge production or a social construct negotiated by conflicting interest groups but as a cultural practice that is complex, heterogeneous, and an important and integral part of our ensemble of cultural practices. It entails many different agents, such as concepts, machines, humans, and animals that produce meanings and thereby maintain or refigure cultural boundaries. As a consequence, cultural studies of technoscience not only analyze material and social technologies but also visual and semiotic ones (Haraway 1991 [1985]; and McNeil and Franklin 1991).

Until recently, cultural studies of technoscience, with its feminist and Marxist traditions, has been widely ignored at least by the mainstream of philosophy, sociology, and history of science and technology. Initiated by cultural and media studies, the discussions on technology as media (which draw heavily on cultural studies) and of technoscience as cultural and everyday practice is broadening and adopted in many different discourses. The role of media and visualization practices for the production of technoscientific knowledge is rethought with regard to new computer-generated images (Heintz and Huber 2001; Nordmann 2007a; and Valerie Hanson's and Angela Krewani's chapters in this edited volume) as well as their close affinity between technoscientific research and artistic practices. Another case in point would be the radical epistemological and ontological shifts in technoscience, moving from representation to performativity, from determinism to unpredictability, from causality to nonlinearity thereby abandoning traditional conceptions of objectivity and subject-object relations (Law and Urry 2003; Pickering 2002; see also Martin Carrier's and Hans Radder's chapters in this edited volume).

The cultural turn of technoscience is indicated by new media of persuasion such as immersion, the new affinity of technoscience toward art as well as the self-understanding of scientists not only as entrepreneurs but also "closet artists" (Risan 1996). Science and technology—that is, technoscience—is no longer mainly about representing the laws of nature and intervening in its processes, but mostly about (re)shaping new and hybrid worlds from a constructivist viewpoint (Weber 2003). This cultural turn is encouraged by technosciences' new epistemologies and ontologies that interpret our world as our product (see Carrier in this edited volume; Haraway 1992), where nature itself becomes an entrepreneur and engineer (Haraway 1997). In this frame of thought, geneticists, nanotechnologists, brain researchers, or roboticists are perceived as technoscientists who mainly support, improve, and perfect nature. We find a new openness to interdisciplinarity, as we know it from the prime time of cybernetics in the 1940s and 1950s, to foster innovation and problem solving by integrating the humanities and social sciences as a promising resource for the undertakings of an entrepreneurial technoscience.

The emergence of new everyday technologies like the PC or the cell phone, possibilities of enhancing bodily functions (for example, prenatal testing, in vitro fertilization, or Viagra), and the increasing integration of human and technological systems, the orientation of technoscience toward application from the 1980s on (Forman 2007), as well as the more intimate relation between science and society after 1989 (Gibbons et al. 1994) has led to a growing public interest in technoscience. Therefore, many technoscientists and science managers invest increasingly in the popularization of technosciences not only to demonstrate the innovative character of one's work in the face of limited societal resources and succeed in the intensified competition for research funds, but also to demonstrate the usefulness of technosciences' endeavors for the public. Aspects of personal service and entertainment are especially coming to the foreground. In such a sociohistorical context, the performance of new robotics with anthropomorphic and zoomorphic artifacts (Aibo, Pleo, Pino, and so on) becomes a promising technoscience that is on the one side consumer-oriented high tech but at the same time inscribes itself into popular culture, promoting the fun part through a very broad variety of entertainment and edutainment applications—without losing the aspect of innovative, high-tech masculine breakthrough work.

Machines of Wonder and Curiosity

From a historical perspective the production of entertaining human- and animal-like machines to rouse feelings of wonder and amazement in their spectators is not new. The automata of ancient times and in the sixteenth and seventeenth centuries were built as media of illusion, as curiosa, as fascinating and entertaining entities that had at least partly also symbolic function (Karafyllis 2004) (figure 13.2).

In contrast to this, the automata of the eighteenth and nineteenth centuries were used as models of perception for scientific research. With the abandoning of the mechanical model of the organic at the end of the nineteenth century and the emergence of its biocybernetic concept in the twentieth century, which shifted the translation between humans and machines to the *systemic and micro-level*, the construction of humanoids was no longer interesting from a scientific perspective. Beginning in the 1950s, the credo of robotics was to develop *functionalist* machines as efficient and optimal problem solvers. And while humanoids as talking, identity-challenging machines were the frequent stars in science fiction (SF) film and literature in the twentieth century, humanoid robots did not play a role in Western academic research until the 1980s.[1] Humanoid robots from science fiction films and literature served as a resource

Figure 13.2. In 1737 the French mechanic and inventor of the mechanical loom, Jacques de Vaucanson, built a mechanical duck that could drink, swim, and excrete.

of inspiration for many roboticists but were not themselves a central object of research.[2] In the past decades this old "division of labor" seems to be chang-ing—and this change is also related to one in research paradigms in robotics.

From the Manipulation of Things toward Personal Services of Well-Being

In traditional industrial robotics, robots were understood as multifunctional manipulators and problem solvers. According to the functionality of these robots, the relation between the roboticist-expert and the machine was mod-eled as a master-slave relation: the engineer directs the actions of the machine, while the latter carries them out. This traditional understanding of robotics is visible in the (still valid) robot definition of the Verein Deutscher Ingenieure (VDI, translated as the Association of German Engineers) guideline for indus-trial robots from 1990: "A robot is a free and re-programmable multifunctional manipulator with at least three independent axles, *to move materials, parts, tools, or special machines on programmed, variable tracks to accomplish various tasks*" (qtd. in Christaller et al. 2001, 18; my translation and emphasis).

But in the late 1980s and the 1990s a radical shift emerged in robotics toward biology and the neurosciences. While traditional robotics was mainly oriented toward mathematics and formal logic, engineers now started to use biological principles to build more intelligent machines. The new approach stressed the importance of robustness, autonomy, embodiment, and situatedness for the creation of intelligent systems.[3] These new systems were often shaped zoomorphic, for example, as snakes, bugs, or ants. This development laid ground for the emergence of "social" robotics or HRI in the 1990s, which builds not zoomorphic but humanoid prototypes with features such as facial expressions, gestures, emotions, "natural" humanlike ways of communication including turn-taking (see, for example, Breazeal 2003).

Today's definition of a service robot is quite different from the traditional one of industrial robotics. The United Nations defined a service robot in the following way: "A robot which operates semi or fully autonomously *to perform services useful to the well being of humans* and equipment, excluding manufacturing operations" (qtd. in Euron and IFR 2004, 1; my emphasis). In HRI the robot is conceptualized as an infant that needs to be cared for and educated. At least the strong approach in HRI follows theories of developmental psychology and uses analogies between robot and child development, searching for more effective ways to build intelligent robots. They want to "evolve" robots that can be educated and shaped as humanoid "cognitive companions."

To give an example of this approach, the European Union's research project Cogniron (2006–9)—funded with 6.5 million Euros by the European Community and Switzerland—describes its research goals in the following way: "The project will develop methods and technologies for the construction of such cognitive robots able to evolve and grow their capacities in close interaction with humans in an open-ended fashion. *The robot is not only considered as a ready-made device but as an artificial creature, which improves its capabilities in a continuous process of acquiring new knowledge and skills. . . . The design of cognitive functions of this artificial creature and the study and development of the continuous learning, training and education process in the course of which it will mature to a true companion, are the central research themes of the project."*[4]

Notwithstanding the fact that EU research policy fosters research that relies on the rhetoric of breakthrough and radical innovation beyond incremental research (Nordmann 2007b)—and thereby supports stories of salvation, wonder, and curiosity—it is nevertheless amazing how the monstrous repetitive manipulators and industrial problem-solving machines have "evolved" to babylike, self-learning artificial creatures of HRI that need to be educated to serve as one's own true companion and that are supposed to support the indi-

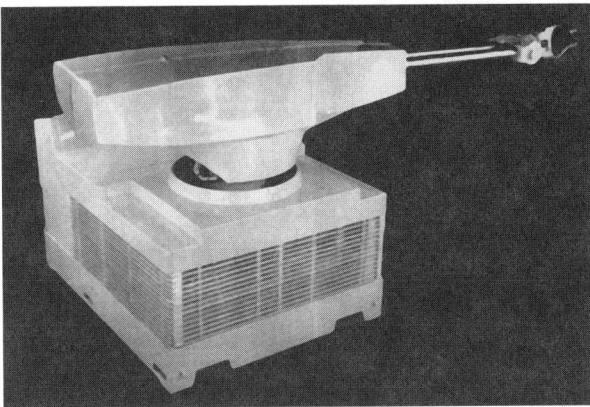

Figure 13.3. Unimate, the first industrial robot deployed at the assembly line of General Motors in 1961.

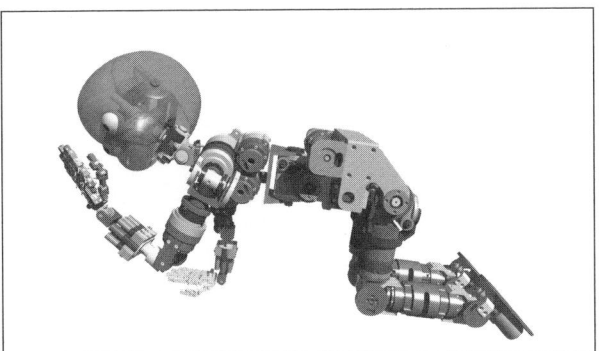

Figure 13.4. Model of iCub. iCub is a humanoid developed by the EU project Cogniron, which was meant to be not only an artificial creature but a "true companion." *Source*: Prof. Giulio Sandini.

vidual human user in the long run. Anthropomorphic, emotional, and social artifacts, once the traditional and exclusive domain of the popular culture of science fiction, are now developed for everyday users. Robotics has shifted from manipulating things and moving materials to providing service gadgets for the well-being of humans, from the mastery of nature toward personal comfort and sociopsychological and therapeutic services, from industrial production to service products and popular culture—a shift that is accompanied by epistemological and ontological changes and a move from infrastructural technologies toward the production of consumer technologies (figures 13.3 and 13.4).

From Biologically Inspired Machines Toward Hyperrealistic Humanoids

But while HRI could be interpreted as the logical precursor of biological-inspired robotics, one should not overlook the epistemological difference. The older biomimetic approach is a biologically inspired approach where biological mechanisms and principles have mainly the function of inspiration and are not copied in a 1:1 fashion, because many organisms are only semioptimal for the task an artificial agent is supposed to fulfill; significant differences are seen between the logic of nature and that of technology. For example, the development of planes was inspired by mechanisms in birds' flying behavior but they were not copied precisely. Airplanes do not flap their wings.

In HRI we find a branch of humanoid robotics that focuses on the construction of (hyper)realistic 1:1 copies of humans. Some engage in copying the bodily shape of humans as perfectly as possible.[5] Others develop mechanisms to copy human facial expressions, gestures, and "mechanisms" of social interaction (e.g., turn-taking) as "natural" as possible. Now humanoids are no more the subject of science fiction alone. Since a few years ago, aesthetic and physical features of so-called social robots have become an intrinsic and central part of the roboticists' work in HRI. Many of its researchers believe the claim by Byron Reeves and Clifford Nass (1996) of a natural tendency of humans to anthropomorphize computers. Along that line, many roboticists make the questionable and unproven assumption that humanoid robots would make a perfect social interface. Humanoids are supposed to help encourage an increased emotional bonding toward machines and the immersion of users into human-robot interaction. Another unproven claim is that mobile humanoid robots will be more easily accepted because they might be perceived as less threatening than functional designed machines (Kiesler and Hinds 2004).

These claims leave open the question of why machines need a humanoid shape to support the interest or even bonding with machines. We already anthropomorphized computers when they still looked like boring tin boxes. Obviously, anthropomorphization is not dependent on machine design. If there is a "natural" tendency to anthropomorphize computers, it is grounded in a human tendency to ascribe computers identity or personality. The "unpredictable" behavior of computers and especially of "self-learning" and "autonomous" machines might support these processes, but this behavior is not limited to the field of computers and robots—think, for example, of the anthropomorphization of cars. At the same time, the applications areas for humanoids are only described very vaguely. In the report of the International Federation of Robotics, humanoids are praised on the one hand as "the ultimate human machine"

Figure 13.5. *(top, left)* **Robosapien is a humanoid robot advertised as "multifunctional, thinking, feeling robot with attitude!"** *Source*: Sebastian Budich and Sablon Germany GmbH / WowWee Group Limited.

Figure 13.6. *(bottom, left)* **Pleo, the robotic baby dinosaur.** *Source*: Derek Paul Dotson from Innovo Labs / COO.

Figure 13.7. *(right)* **Pino, the "interactive robot friend."** *Source*: Nobuko Imanishi and ZMP Inc.

because of their adaptability to human environments, but on the other hand the report states very vaguely that they will be "able to potentially cover a wide variety of jobs" (Hägele 2006, 424). And the text proceeds: "However, technological and manufacturing challenges still remain to be solved until reliable machines can be produced in significantly high quantities" (Hägele 2006, 424).

Up to now, the most important field of application for personal service ro-

Figure 13.8. (left) **Aibo is an autonomous robot and interactive pet.** *Source*: Désirée Kuhn and Sony Deutschland GmbH.

Figure 13.9. (below) **Paro, the "therapy" robot for elderly people.** *Source*: Takanori Shibata PhD and AIST Japan.

bots and especially humanoids are the entertainment and edutainment industry, like Robosapien, Pleo, Pino, and other toy robots. In 2005, 1 million units in total were sold worldwide. For 2006 through 2009 the numbers are estimated about 5.5 million (Hägele 2006). The sex industry might also be an important application area. In Korea we already find robot hostess services, but numbers of sold units are not available (figures 13.5, 13.6, and 13.7).

In addition, the research field of humanoid robotics struggles with the problem of the "uncanny valley." This thesis postulates that only vaguely humanlike machines are willingly accepted, while hyperrealist humanoid machines cause not only uncanny feelings but the rejection of these robots. Cartoonlike or zoomorphic robots for emotional bonding and immersion would do as well—as we can see with the successful robodog Aibo or the baby seal robot Paro (figures 13.8 and 13.9).

Having these difficulties and theoretical shortcomings in mind, the question is: Why are humanoid robots so attractive for research and development? What might be the most promising about humanoids is that they serve as promotion devices for consumer technology in the context of (life-style) design. For example, the already mentioned Report on World Robotics describes the EXPO 2005 in Aichi as "[a] highlight for humanoid robot technology" because "it staged a robot week in June of that year with a large variety of demonstrators and prototype robots from companies, research institutes and universities. *Humanoid robots have been a highlight in particular as these were displayed in impressive shows and settings*" (Hägele 2006, 423; emphasis added). Maybe this is the most convincing explanation for the huge interest in and the funding of the quite expensive but not very productive field of humanoid robotics.

Design, Affect, and Consumer Technologies of Personal Service Robotics

Since the 1970s, the borders between technoscience and society have become increasingly permeable.[6] The "sacramentalism of science" (Bensaude-Vincent 2001, 108) in the twentieth century permanently lost ground with the growing criticism of such social movements as the ecology movement, the women's health movement, as well as antinuclear activism, the end of the cold war, and the growing public awareness of the ambiguous role of science in the production of crisis (Michael 1998). Science and technology became increasingly a contested terrain with rising claims for the justification of research funds and policy.

At the same time, the orientation of technoscience toward application, the higher dependency of universities on government and international funding, and the pressure to participate in the production of key technologies calls for the efficient promotion of one's research in the mass media. "Social," emotional, and humanoid machines endowed with a rich techno-imaginary are quite attractive and effective tools in the media economy of attention. Smiling humanoids climbing stairs, dancing with human partners, or talking with curious visitors easily find their way into the latest news. They ensure public attention via the mass media, and this might compensate for poor application areas. On the other hand, personal (humanoid) service robots fit very well into the demands of a growing service economy that does concentrate less on infrastructure technologies but on individual consumer technologies for the (white) middle-class techno-educated users (figures 13.10 and 13.11).

The general interest in humanoid robots and their function as eye-catchers and high-tech demonstrators in conferences, fairs, and media events has at least four aspects:

Figure 13.10.
PBDR, the Partner Ballroom Dance Robot for shy guys. *Source*: Prof. Yasuhisa Hirata and the System Robotics Laboratory at the Tohoku University, Nomura Unison Co., Ltd., and TroisO Co., Ltd.

Humanoid robots serve as charming, pleasurable, entertaining, and aesthetic artifacts and as a "living" sign for the high-tech culture his or her owner belongs to. Their "humanness" reawakens the old feeling of magic and wonder—which brings us back to the function of automata as media of illusion, as curiosa, as fascinating and entertaining entities. Personal service robotics invests in the development of individual consumer technologies using immersion, affective bonding, and attractive design to promote caring, assisting, and especially entertaining machines. Therefore, human-robot interaction (in analogy to human-computer interaction) becomes an innovative research field that relates successfully to the aesthetization of everyday life and the requirements of consumer culture. Tomorrow's humanoid robots are already praised as the artifact with an unbelievable wide variety of possible functions. Working properly, they are supposed to be substitutes for human workers in the service economy on a wide range—as industrial robots have in the industrial sector from some decades ago until now.

Figure 13.11. Qrio, a humanoid robot and proposed follower of Aibo, also developed by Sony, which was supposed to substitute the Aibo but never did. *Source*: Désirée Kuhn and Sony Deutschland GmbH.

At the same time, the future personal robot—then endowed, it is hoped, with "true intelligence"—also reawakens old colonial and sexist dreams of the always available, obedient servant or submissive housewife who never makes any demands but cares for the master and anticipates his wishes. But imaginations about personal servants also carry the promise of freeing one from everyday duties in the home and the extensive care for the owner of the personal robot. Also, human-robot interaction inscribes itself efficiently into today's Western popular culture with cute artifacts such as Aibo, Asimo, Pleo, or Pino while demonstrating its high-tech capabilities.

"Technoscience Is Fun!"

In terms of science communication, humanoid robots are very rich resources playing not only on the techno-imaginary of science fiction pop culture but also on the tradition of artifacts as source of wonder, curiosity, and amazement. The idea to entangle users in HRI instead of enabling them to control the robot, as well as the importance of humanoid aesthetics, results from the changing relationship of technoscience and media, of today's technosciences, and popular culture. This strategy of immersion is supplemented with one of edutainment and partial participation.

Edutainment and toy robots, such as Aibo or Pino as well as other humanoid or zoomorphic artifacts, provide good opportunities not only to involve laypeople in one's technoscience but also to attract future researchers. Think of the latest dream of the robotics community—a soccer game between humans and robots in 2050. Science festivals like the annual robot soccer competition RoboCup, science fairs, courses in "educational robotics," science museums, and artists-in-residence are other popular strategies and events to attract public attention and to heighten the performativity of one's research and latest artifacts. Robotics—like AI before it—is quite effective in marketing its latest achievements and visions. Edutainment as well as sweet-looking and friendly artifacts help to reshape robotics from a boring engineering discipline with its ugly machines toward a sexy, challenging, and interesting field with friendly, social, intriguing, and—if desired—sexy and unpredictable artifacts. Robotics equips itself with a high entertainment value and stages itself as the result of hard and innovative high-tech breakthrough work at the same time.

This way of self-representation of emerging technosciences can also be seen as a new style in science communication. Science communication transforms from a one-way channel used by an authoritative expert culture to communicate their research findings to the public toward a technoscience communication that builds on edutainment and fun. In a time when citizens are often equated with consumers (Michael 1998), technoscience is presented as fun and edutainment and reshapes itself as user-friendly and open to the public.

Technoscience Engineering Philosophy

At the same time, the old authoritative style of science communication is not totally abandoned; rather, it is reconfigured in an interesting way. While robots are transformed from problem solvers toward social, sexy, and entertaining machines, engineers transform from producers of problem solvers to applied philosophers. As neuroscientists or geneticists were before them, roboticists are now frequently consulted by the media not only to give their technical expertise but to inform us about proper conceptions of humanness and the human condition.[7] Interestingly, although science must justify its research and its work must be application-oriented, we have at the same time a trend toward a renaturalization of body, mind, and society. In search of new certainties, not only the expert opinion of technoscientists but also their "philosophical" judgment is increasingly gaining importance. Technoscientists are increasingly staged as public "philosophers" not only rebuilding bodies and artifacts but consciously and skilfully reshaping our human nature. Roboticists themselves claim that research on robots is helpful to learn about human physiology and nature. Humanoids are interpreted as necessary tools to understand the functioning of humans.[8] And this might be more than a justification for expensive research where applications are still missing.

Humans become the inspiration and model for machine design, and in a feedback loop humans are interpreted in the light of machines. Research programs are "positing the intelligent machine as the appropriate standard by which humans should understand themselves. No longer the measure of all things, man now forms a dyad with the intelligent machine such that man and machine are the measure of each other. We do not need to wait for the future to see the impact that the evolution of intelligent machines has on our understandings of human beings. It is already here, already shaping our notions of the human through similarity and contrast, already becoming the looming feature in the evolutionary landscape against which our fitness is measured. The future echoes through our present so persistently that it is not merely a metaphor to say the future has arrived before it has begun. When we compute the human, the conclusion that human beings cannot be adequately understood without ranging them alongside the intelligent machine has already been built into the very language we use" (Hayles 2003, 116).

It is not surprising that our future copies are so well functioning in the economy of attention. From a functionalist perspective it would be more interesting to build functional robots that solve problems that are unsolvable by humans—for example, by using infrared sensors or laser scanners. In the field of humanoid robots useful functions in the sense of problem-solving capabilities seem to become secondary while the field adapts to the demands of the service economy and consumer culture focusing on entertaining and edutaining and media-effective artifacts. At the same time, it plays on techno-imaginaries from technology history and science fiction pop culture. The integration of technoscience into popular culture becomes a higher priority as technoscience communication is reconfigured as edutainment, fun, and philosophy.

At a historical moment when the relation between science and society opens up and a renewed discourse between public and professional science seems to be possible (Bensaude-Vincent 2001), I wonder if there could be alternatives beyond a hierarchical organized technoscience communication from technoscientists toward the public. A popular but not popularized technoscience embedded in a public discourse could be such an alternative. Educational robotics would be a perfectly suited field for such an enterprise instead of humanoid robotics. But instead, technoscience communication lingers between (re)presenting technoscience as edutainment, fun, the foundation for philosophical expertise, and a source for magic and wonder.

On the one hand, the sacramentalism of science is increasingly deconstructed and a more open and user-friendly marketing of technoscience can be observed. But this new discourse focuses on pleasure, emotion, and immersion, thereby following the demands of today's consumer culture and the stereotypical and more simpleminded aspects of old techno-imaginaries. Technoscience is either reduced to fun or serves as truth discourse on the human condition. I wonder whether at least the fun part—in a less simplified version—could be a starting point to reinvent a new culture of public science. In this new technoscience culture we would not be immersed in old-fashioned stereotypes of master-servant, infant-caregiver relations; instead, we would be immersed in dreams of overall wellness but revive a public culture of participation and the search for more liveable technoworlds.

NOTES

I am very grateful to Alfred Nordmann, Hans Radder, Gregor Schiemann, and all the other members of the ZiF research group Science in the Context of Application (2006–7) at the University Bielefeld as well as to the anonymous reviewers of this chapter for many helpful comments. Many thanks to all the researchers and companies who allowed us to reproduce their pictures. Furthermore, I want to thank my assistant Andreas Weich for investigating the manifold copyright issues; Alfred for supporting my point that sometimes pictures are not only helpful but necessary to make one's argument; Hans for his patience; and the University of Pittsburg Press—without them I would not have been able to include this bestiary of robots in this chapter. The chapter epigraphs are from Michael 1998, 321; and Haraway 1997, 66.

1. Related to its historical and cultural situation, we already find research in humanoids in the 1970s in Japan.
2. Roboticists are also working in animatronics for the film and theme park industry.
3. See Weber 2008. Some cybernetic researchers, such as Grey Walter, built on self-learning, autonomy, and robustness in the 1940s but were marginalized from the 1960s on with the growing dominance of symbol-processing AI.
4. Cogniron, "The Cognitive Robot Companion, FP6-IST-002020," online at http://www.cogniron.org/InShort.php (emphasis mine).
5. See the pictures of the Geminoid project of Prof. Ishiguro in which he and his research group developed a robot that aims to look like a twin of the professor or the shaping of humanoid robot heads like Einstein, Julio, or Alice by Hansonrobotics (see http://www.irc.atr.co.jp and http://www.hansonrobotics.com/products.htm). I call these humanoids hyperrealist because of the enormous efforts undertaken to make them more "human" than a human can ever be. It reminds me of the performative and elaborate work of transvestites to stage the perfect woman—thereby being more "female" than a woman can ever be.

6. Bernadette Bensaude-Vincent (2001) gives an illuminating historical overview of the relationship between science and the public. She shows that in the eighteenth century, popular science as a common practice was highly valued, and from an amateur perspective science was often perceived in a broad perspective. In the twentieth century a gap between science and the general public grew in the context of new physics and its radical break with common sense.
7. Many thanks to Alfred Nordmann for mentioning this important aspect.
8. See, for example, the new research Institute for Cognition and Robotics at the Bielefeld University, where sport scientists want to work out strategies of optimal human movement by studying the movements of humanoids.

REFERENCES

Bensaude-Vincent, Bernadette. 2001. *A Genealogy of the Increasing Gap between Science and the Public.* Paris: University of Paris.

Breazeal, C. 2003. "Emotion and Sociable Humanoid Robots." *International Journal of Human-Computer Studies* 59, nos. 1–2 (July 2003): 119–55.

Christaller, Thomas, Michael Decker, Joachim-Michael Gilsbach, Gerd Hirzinger, Karl Lauterbach, Erich Schweighofer, Gerhard Schweitzer, and Dieter Sturma. 2001. *Robotik: Perspektiven für menschliches Handeln in der zukünftigen Gesellschaft.* Berlin: Springer.

European Robotics Forum (EURON) / International Federation of Robotics (IFR), Martin Hägele, and Hendrik I. Christensen, eds. 2004. "European Service Robotics: A White Paper on the Status and Opportunities of European Service Robotics." Online at http://www.cas.kth.se/euron/euron-deliverables/ka4-4-white-paper.pdf.

Forman, Paul. 2007. "The Primacy of Science in Modernity, of Technology in Postmodernity, and of Ideology in the History of Technology." *History and Technology* 23, nos. 1–2: 1–152.

Gibbons, Michael, Camille Limoges, Helga Nowotny, Simon Schwartzman, Peter Scott, and Martin Trow. 1994. *The New Production of Knowledge: The Dynamics of Science and Research in Contemporary Societies.* London: Sage.

Hägele, Martin. 2006. "Service Robotics." In *World Robotics 2006—Statistics: Statistics, Market Analysis, Forecasts, Case Studies, and Profitability of Robot Investment*, edited by International Federation of Robotics and Statistical Department Robotics and Automation Association (Verband Deutscher Maschinen- und Anlagenbau e.V. [VDMA], Frankfurt am Main and Fraunhofer-Institut für Produktionstechnik und Automatisierung [IPA] Stuttgart), 377–446. Frankfurt.

Haraway, Donna J. 1991 [1985]. "Manifesto for Cyborgs: Science, Technology, and Socialist Feminism in the Late Twentieth Century." In *Simians, Cyborgs, and Women: The Reinvention of Nature*, edited by Donna Haraway. London: Routledge. Originally printed in *Socialist Review* 80 (1985): 65–108.

———. 1992. "The Promises of Monsters: A Regenerative Politics for Inappropriate/d Others." In *Cultural Studies*, edited by L. Grossberg, C. Nelson, and P. A. Treichler, 295–337. New York.

———. 1997. *Modest_Witness@Second_Millennium: FemaleMan_Meets_OncoMouse: Feminism and Technoscience.* New York.

Hayles, N. Katherine. 2003. "Computing the Human." In *Turbulente Körper, soziale Maschinen: Feministische Studien zur Technowissenschaftskultur,* edited by J. Weber and C. Bath. Opladen: Leske and Budrich.

Heintz, Bettina, and Jörg Huber. 2001. *Mit dem Auge denken: Strategien der Sichtbarmachung in wissenschaftlichen und virtuellen Welten.* Zürich: Edition Voldemeer.

Karafyllis, Nicole 2004. "Bewegtes Leben in der frühen Neuzeit: Automaten und ihre Antriebe als Medien des Lebens." In *Technik in der frühen Neuzeit—Schrittmacher der europäischen Moderne,* edited by Gisela Engel and Nicole Karafyllis, 295–335. Frankfurt: Klostermann.

Kiesler, Sarah, and Pamela Hinds, eds. 2004. *Human-Computer Interaction* 19, no. 1–2 (special issue).

Law, John, and John Urry. 2003. "Enacting the Social." Department of Sociology and the Centre for Science Studies, Lancaster University. Online at http://www.comp. lancs.ac.uk/sociology/papers/Law-Urry-Enacting-the-Social.pdf.

McNeil, Maureen, and Sarah Franklin. 1991. "Science and Technology: Questions for Cultural Studies and Feminism." In *Off-Centre: Feminism and Cultural Studies,* edited by Sarah Franklin, Celia Lury, and Judith Stacey, 129–46. New York: HarperCollins.

Michael, Mike. 1998. "Between Citizen and Consumer: Multiplying the Meanings of the Public Understanding of Science." *Public Understanding of Science* 7: 313–27.

Nordmann, Alfred. 2007a. "Beholding the Objects of Science." Unpublished paper.

———. 2007b. "Ignorance at the Heart of Science: Incredible Narratives on Brain-Machine Interfaces." Technische Universität Darmstadt. Online at http://www. uni-bielefeld.de/ZIF/FG/2006Application/PDF/Nordmann_essay.pdf.

Pickering, Andrew. 2002. "Cybernetics and the Mangle: Ashby, Beer, and Pask." *Social Studies of Science* 32, no. 3 (June 2002): 413–37.

Reeves, Byron, and Clifford Nass. 1996. *The Media Equation: How People Treat Computers, Television, and New Media Like Real People and Places.* Cambridge: Cambridge University Press.

Reinel, Birgit. 1999. "Reflections on Cultural Studies of Technoscience." *European Journal of Cultural Studies* 2, no. 2: 163–89.

Risan, Lars. 1996. "Artificial Life: A Technoscience Leaving Modernity?" Thesis in anthropology at the University of Oslo. Online at http://www.anthrobase.com/ Txt/R/Risan_L_05.htm.

Weber, Jutta 2003. *Umkämpfte Bedeutungen: Naturkonzepte im Zeitalter der Technoscience.* Frankfurt am Main: Campus.

———. 2005. "Helpless Machines and True Loving Caregivers: A Feminist Critique of Recent Trends in Human-Robot Interaction." *Journal of Information, Communication, and Ethics in Society* 3, no. 4 (2005): 209–18.

———. 2008. "Human-Robot Interaction." In *Handbook of Research on Computer-Mediated Communication,* edited by Sigrid Kelsey and Kirk St. Amant, 855–67. Hershey, Pa.: Idea Group Publisher.

The Good Old Days
Medical Research
Then and Now

JAMES ROBERT BROWN

THERE WAS NEVER A GOLDEN AGE when medical research was all sweetness and light. Only fools could think otherwise. Yet, in many respects, former times were better times. The quality of research has suffered from the assault inflicted upon it from corporate interests and their scientific hirelings who are more concerned with mammon than medicine. This is a major change and it is not for the better. It is partly the result of significant discoveries, partly of methodological innovation, and partly of social and institutional change. Inevitably, some will see this as the natural evolution of science, technology, and society. And yet it is anything but natural, if by natural we mean something to be accepted because of its inevitability. I will try to say why.

Is science today *continuous* with the practices of the past, or is there a sharp *break* with what went before? In addressing the continuity-versus-break issue, the editors of this collection suggest three different ways to see the relation between so-called pure science and technological applications. Much of the debate will turn on this issue.

- Scientific research creates new technical capabilities that are then developed in engineering contexts (with more or less prominent feedback processes—for example, Heinrich Hertz and the radio).

- Technological innovation gets ahead of scientific understanding and prompts research activity to attain comprehensive understanding of its basic principles (be it to better manage the technology, be it to gain fundamental insight from technologically produced phenomena—for example, thermodynamics and the steam engine, plasma physics).

- Piecemeal research activities to manage complexity of sociotechnical systems with no expectation of comprehensive understanding (nanotoxicology, foresight knowledge, commissioned explorative research to support specific public decisions, and so on).

This is fine as far as it goes, but the list of possibilities is not exhaustive. The three options of this volume's editors can be put simply: science helps technology, or technology helps science, or science and technology work more or less independently. The missing relation from the list of possibilities is: *technological aims subvert good science.* It is certainly not inevitable, but it is exemplified all too often, especially in medical research. The reason has nothing to do with the intrinsic nature of technology itself but with the financial goals of those who are developing the technology. A brief comparison of past and present should make this clear.

The Past: Water Pills

If I were trying to make a case for the good old days, it would be rhetorically effective to cite a few dramatic examples, such as the discovery of the polio vaccine. When asked if he would be patenting his discovery, Jonas Salk remarked, "Could one patent the Sun?" Though some today would like to do just that and are pushing the patenting possibilities as far as they can, Salk exemplified the romantic image of science that inspired some of us as children, an image that many would like to cling to still. Cherish it though we may, we know better. Most science is not and never was like this; it is filled with drudge work, piecemeal advances, professional jealousies, and other decidedly unromantic aspects. Nevertheless, there were real accomplishments, such as Salk's, and the quality of our lives has improved greatly because of them. One of these accomplishments was the discovery of diuretics for reducing blood pressure. It was not the breathtaking stuff of Hollywood movies, but spectacular nonetheless for all the lives it has saved over the years.

Hypertension is unlike other diseases in that its symptoms are usually unnoticed—one could have very high blood pressure and feel perfectly fine. In the United States an estimated forty million people have hypertension; more than half of these are treated with some medication. Early attempts to control it occurred about midcentury. The development of diuretics at this time, the late 1940s, provided a relatively safe and effective solution. These are commonly known as water pills. Within a few years their enormous potential was recognized. In one test carried out on U.S. military veterans, two of seventy who were treated suffered strokes, compared with twenty-seven of seventy who were untreated (Hamdy 2001). Diuretics were followed in the 1960s, first with the development of beta-blockers, then with a variety of other drugs.

An interesting feature of beta-blockers is that they were designed, not accidentally discovered. This was perhaps an important development in promoting the pharmaceutical industry's interest in highly focused research and specifically in "designer drugs" (see Matthias 2005). It certainly proved profitable. In the intervening years, there have been a significant number of products developed to control hypertension, mainly by for-profit corporations. These have been widely advertised to the public, widely promoted to doctors, widely prescribed, and widely used. We might wonder: In what ways are these commercial products improvements over the older diuretics? Not much, it turns out.

The Antihypertensive and Lipid-Lowering Treatment to Prevent Heart Attack Trial (known as the ALLHAT study) published in the *Journal of the American Medical Association* (*JAMA*) in 2002 answered the question in a dramatic way. I quote at length from a laudatory editorial by Lawrence Appel that was published in the same journal issue. Appel nicely summarizes the study and its significance and, above all, he is a neutral authority on these matters. I would draw attention to three key passages, which I have italicized. In what follows, chlorthalidone (thiazide) is the diuretic. Appel begins with a summary of the remarkable results:

> Quite simply, the Antihypertensive and Lipidlowering Treatment to Prevent Heart Attack Trial (ALLHAT) is one of the most important trials of antihypertensive therapy. For decades, experts have passionately debated which class of drugs should be initial therapy for hypertension. Resolution of this issue, which has enormous clinical, public health, and economic implications, comes at a time of intense pressure to reduce health care costs while improving clinical outcomes. In this setting, *the ALLHAT results, reported in this issue of [JAMA], are particularly noteworthy, because there is no cost-quality tradeoff; the most effective therapy was also the least expensive.* (ALLHAT 2002, 2981, emphasis mine)

Appel next describes the methodology of the trials, which I will skip. Then comes the principal outcome:

> *The major finding of ALLHAT was a striking and unequivocal null result, namely, that the occurrence of coronary heart disease death and nonfatal myocardial infarc-*

tion was virtually identical in the amlodipine, lisinopril, and chlorthalidone groups.
The data effectively rule out a 10 percent or more difference between chlorthalidone and each of the other therapies. However, for certain secondary outcomes, there were apparent differences, some of which were anticipated. Chlorthalidone was superior to amlodipine in preventing heart failure. This finding is consistent with trends observed in other trials. (Appel 2002, 3040; emphasis mine)

Appel fills in a bit more detail, contrasting the ALLHAT study with others. I will skip this and go to the final comments on the social-political consequences.

A logical strategy that incorporates these low-cost agents may differ from those that are more popular, but contemporary strategies may be somewhat artificial because of the heavy influence of marketing that preferentially leads to use of expensive medications. In short, physicians have the means to effectively control BP [blood pressure] *with inexpensive medications, even among patients who require multiple drugs.* (Appel 2002, 3041; emphasis mine)

There are several things to note. First, the older diuretic medication proved itself to be equal or better than newer alternatives in a number of ways, including effectiveness at reducing blood pressure, relative lack of side effects, and cheaper cost. So one need not trade effectiveness for lower cost or reduced side effects. Second, the current extensive use of more costly drugs is, as Appel remarks, the likely result of large-scale advertizing, not merit. And third, though not mentioned by Appel, the ALLHAT study was not funded by for-profit corporations, but by nonprofit foundations and the U.S. National Institutes of Health. The rule of thumb "Follow the money" is as relevant in medical research as it is anywhere else.

I said at the outset that I would not wish to give the impression that medical research in the past was crowded with triumphs and that everything has been downhill since. But the ALLHAT study does reveal something striking—namely, that frequently there are economic forces currently at work that undermine, misdirect, and even corrupt our best research efforts. Half a century ago, medical research was less likely to be tainted, because commercial interests were quite insignificant compared with today. There are examples galore to illustrate the current lamentable state.

The Present: Follow the Money

Richard Davidson (1986) found that in his study of 107 published papers that compared rival drugs, the product produced by the sponsor of the research was found to be superior in every single case. Need one be a mindless cynic to think this is not a mere coincidence? M. Friedberg and coauthors (1999) found that only 5 percent of published reports on new drugs that were sponsored by the developing company gave unfavorable assessments. By contrast, 38 percent of published reports were not favorable when the investigation of the same drugs was sponsored by an independent source. H. Stelfox (1998) studied seventy articles on calcium-channel blockers. The articles in question were judged as *favorable, neutral,* or *critical.* Their finding was that 96 percent of the authors of *favorable* articles had financial ties with a manufacturer of calcium-channel blockers; 60 percent of the authors of neutral articles had such ties; and only 37 percent of authors of unfavorable articles had financial ties. Incidentally, in only two of the seventy published articles was the financial connection revealed. Given these studies and others like them, it is perfectly reasonable to say that there is a pattern of corruption in medical research thanks to the commercialization of the field. We should be worried about who is funding the research. Let's look at a few more examples.

Celebrex, which is used in the treatment of arthritis, was the subject of a yearlong study sponsored by its maker, Paramacia (now owned by Pfizer). The study purported to show that Celebrex caused fewer side effects than older arthritis drugs. The results were published in *JAMA* along with a favorable editorial. It later turned out that the encouraging results were based on the first six months of the clinical trial. When the whole study was considered, Celebrex held no advantage over older and cheaper drugs. On learning this, the author of the favorable editorial was furious and remarked on "a level of trust that was, perhaps, broken" (qtd. in Angell 2004, 109).

Selective serotonin re-uptake inhibitors, known simply as SSRIs, have been central in the new generation of antidepressants. Prozac is the most famous of these. There are several drugs in the SSRI class, including fluoxetine (Prozac), paroxetine (Paxil, Seroxat), sertraline (Zoloft), and others. They are often described as miracle drugs, bringing significant relief to millions of depressed people. The basis for the claim of miraculous results is a large number of clinical trials, but closer inspection tells a different story. There are two related issues, both connected to nonreporting of evidence from clinical trials. C. J. Whittington and coauthors (2004) reviewed published and unpublished data on SSRIs and compared the results. To call the findings "disturbing" would be an understatement. The result was favorable to fluoxetine, but not to the others. They summarized their findings as follows: "Data for two published trials suggest that fluoxetine has a favorable risk-benefit profile, and unpublished data lend support to this finding. Published results from one trial of paroxetine and two trials of sertraline suggest equivocal or weak positive risk-benefit profiles.

However, in both cases, addition of unpublished data indicates that risks outweigh benefits. Data from unpublished trials of citalopram and venlafaxine show unfavorable risk-benefit profiles" (Whittington et al. 2004, 1341).

The related second point is illustrated in a GlaxoSmithKline internal document that was recently revealed in the *Canadian Medical Association Journal*. They were applying to regulatory authorities for a label change approving paroxetine (Seroxat) to treat pediatric depression. The document noted that the evidence from trials were "insufficiently robust" but further remarked: "It would be commercially unacceptable to include a statement that efficacy had not been demonstrated, as this would undermine the profile of paroxetine" (qtd. in Kondro and Sibbald 2004, 783). Annual sales of Seroxat at the time were close to five billion dollars. I should also mention, if only in passing, the high number of suicides associated with Prozac and other SSRIs. This was slow to come to light, but is well documented by David Healy (2003).

Even more stunning results have come to light quite recently. In a thorough study of many clinical trials, using data acquired under freedom of information laws, Irving Kirsch and coauthors (2008) determined that commonly used antidepressants are not effective at all for most people. There is some effect on the very severely depressed, but even here the matter is problematic. The vast majority drugs such as Prozac are, it turns out, no better than a placebo. I should mention that the authors report that they received no specific funding for this work. The significance of this study is succinctly described in the "Editor's Summary" following the article:

> These findings suggest that, compared with placebos, the new-generation antidepressants do not produce clinically significant improvements in depression in patients who initially have moderate or even very severe depression, but show significant effects only in the most severely depressed patients. The findings also show that the effect for these patients seems to be due to decreased responsiveness to a placebo, rather than increased responsiveness to medication. Given these results, the researchers conclude that there is little reason to prescribe new-generation antidepressant medications to any but the most severely depressed patients unless alternative treatments have been ineffective. In addition, the finding that extremely depressed patients are less responsive to placebo than less severely depressed patients but have similar responses to antidepressants is a potentially important insight into how patients with depression respond to antidepressants and placebos that should be investigated further. (Kirsch et al. 2008, 268)

This is just a glimpse at some of the many serious problems plaguing current medical research. Very much more could be said (and is said, for example, in Angell 2004, Biddle 2007, and Brown 2008). One might think this is somewhat one-sided, and that a fair account would also show the enormous benefits of market-driven medicine. It is widely thought that intellectual property rights have done much to stimulate major new advances in medical research from which we have all benefited. But what evidence do we have for this belief, aside from the corporate commercials that tell us how wonderful they are? Such evidence as we do have points in the opposite direction. From 1998 to 2002 the U.S. Food and Drug Administration (the FDA) approved 415 new drugs and classified them for their own purposes as follows: 14 percent were new innovations; 9 percent were significantly improved old drugs; and 77 percent were no better than existing drugs.

The last of these are the so-called me too drugs. They are copies of existing drugs (not exact copies, since they have to be different enough to be patentable). The FDA is obliged to grant approval so long as a new drug is "effective"—that is, does better than a placebo in a clinical trial. In 2002 the FDA approved seventy-eight new drugs. Only seven were classified as improvements over older drugs. Not one of these seven was produced by a major U.S. drug company (see Angell 2004, 75).[1] It seems that intellectual property rights are the principal cause of this. Other factors may be at work, too. Perhaps the "low-hanging fruit" has been picked (largely at public expense). New discoveries are generally getting much harder and are only funded by corporate interests. It seems to me that one has to dance around quite a bit to make explanations other than corporate interests plausible. In any case, I should mention that the Bush administration, under pressure from pharmaceutical lobbyists, instructed the FDA to terminate their classification.

The Open Future

The past gave us high-quality diuretics to control hypertension: water pills are effective, safe, and cheap. The present has provided a plethora of alternatives, all very well marketed but none so effective, none so safe, and none so cheap. When we look at the many other examples of current medicine I outlined earlier, the problem is glaringly obvious. Corporate interests have corrupted a system of medical research that previously served us well, if imperfectly, and rendered it much less beneficial to humanity's health needs. The future is in our hands and there are different ways we may choose to go. One way is to continue on in much the same direction. By taking this route, we leave the process of discovery and development to the various corporations who are happy to direct medical research. Of course, this comes with a cost—steep royalties, flawed products, and research focused narrowly and exclusively on health solutions that lead to intellectual property rights. This is the way that gives us numerous expensive drugs to control hypertension when cheap, effective, and safe diuretics are available. This is the way that suppressed information from clinical trials so as not to undermine the market for highly profitable drugs. This is the way that promotes pharmaceutical solutions to health problems while ignoring diet, exercise, and environmental concerns, since the latter cannot be patented. Must we make such a dispiriting choice?

There is a more promising alternative, involving a considerable amount of regulation. The idea, obviously, is that with the right rules in place and properly enforced, we can rid ourselves of all the ills of money grubbing. It is by far the most popular approach among those who comment on the current situation. What sort of regulations are possible? Here is a brief sample of plausible candidates:

- Require full disclosure by authors of any financial interest when publishing.
- Require advance registration of any clinical trial.
- Establish an independent agency that designs, conducts, and interprets all clinical trials.
- Require clinical trials to test products against leading alternatives as well as against placebos.
- Disallow finder's fees (including disguised equivalents) for recruiting patients into clinical trials.
- Disallow corporate payments of any sort to medical researchers or practitioners, including "education" trips and so on.
- Disallow any corporate presence in medical schools.
- Disallow public marketing of drugs.

There is much to be said about each of these, but I will refrain, except for a few brief remarks. Much of the current emphasis on regulation is aimed at disclosure. It is thought that if only we know who is in a conflict-of-interest position, all will be well. Though full disclosure is certainly a good thing, it is not a panacea. Much more important is the public control of clinical trials. Publicly funded agencies should be fully responsible for the design, execution, and interpretation of results of any clinical trial. Leaving them, as at present, in the hands of corporations is nothing short of absurd. Clinical trials are the principal source of evidence and they cannot be rechecked the way a mathematical proof can be checked by anyone with the appropriate training. Clinical trials cost hundreds of millions of dollars. They can be carried out only once. They must be carried out by those we know we can trust (see Brown 2010b, for more on this point). As for preventing pharmaceutical companies from influencing (to the point of bribing) doctors and medical students, this must seem obvious to all, and yet the corrupt practice is a commonplace. Advertizing drugs is similarly a bad idea, since the public has no way to evaluate the claims being made, but they will pester doctors for certain treatments whether appropriate or not. Advertizing iPods or the latest fashion in shoes is quite different. People can readily tell if their latest entertainment gadgets work and no great harm is done if they do not. But most of us would be hard-pressed to say whether our medication is all it could be. Advertising blood pressure medicines is geared to royalties; the older diuretics provide none, so they were not advertized at all, in spite of being the best in every sense.

Taking clinical trials out of the hands of corporate researchers and placing them under the control of a neutral public body is more than mere regulation. It is a significant political act, akin to nationalizing the very process of medical research evaluation. And it may amount to the first step in the direction of something considerably more radical than increased regulation. This possibility points us in the direction of the third proposal for the future organization of medical research. The various regulations that have been instituted in recent years have improved the situation considerably, but they have not solved the problems and they seem unlikely to do so. The amounts of money involved in the pharmaceutical business are huge and the corresponding temptations to corruption are correspondingly great. But even if the regulation route were successful in every one of its aims, there would still be a major problem that it cannot address. This is the problem of skewed research. All the money and energy of corporate-funded medical research goes into products that will generate income for the investors. This is what motivates the search for drug solutions to health problems. Diet, exercise, and environmental approaches will be ignored or even actively discouraged, since they cannot lead to patents.

Must we endure such second-rate science and technology? There is an obvious solution to this problem, but it will seem quite radical. The solution is simply this: *remove all patents and intellectual property rights in the area of medical research and make all funding public.*

Is such a future for medical science and technology possible? This proposal to socialize medical research may seem too radical and impractical, but it actually is not, at least not in many societies. In Canada, Europe, and in other countries with socialized medicine, the attitude of the general public is that medicine is one place where the market should not rule. In a context where socialized medicine already exists, a proposal such as mine fits seamlessly into the national health care system, where for-profit medical research will seem as out of place as for-profit medical treatment. Of course, this is not true in the United States, but elsewhere in the industrialized world it is. Consequently, as a policy it should be relatively easy to implement in much of the world: social-ized research goes hand-in-hand with socialized medicine, since it is a natural extension of any national health care system. Even for Americans, it need not be viewed as all that radical and could be seen as a return to pre-1980 medical research, when the Bayh-Dole Act was passed. Until then American medical research was almost entirely publically funded through the National Institutes of Health and various foundations, without intellectual property rights. Those were the good old days and the future could look a lot like them.

A word about patents in other fields. If they should be eliminated in med-ical research, why not everywhere? The answer is simple: each case has to be made on its own merits. Perhaps electronics research is better with patent pro-tection while medicine is better without. The situation seems similar to other sectors of the economy. Education, transportation, and health care seem best run publically, while restaurants and the clothing industry seem best in private hands. We should be cautious or even distrustful of any across-the-board argu-ment and always look to the details of particular situations. At this stage my claim is limited to medical research.

I now return to where I started. The initial question concerned the relation-ship of science and technology. Three models were on offer: (1) science helps technology, (2) technology helps science, or (3) science and technology work independently. I said that there is a missing relationship from the list of pos-sibilities: (4) *technological aims subvert good science.* It should now be clear what I mean. But it should also be clear that there are some further distinctions to be made. We shouldn't speak of science as a unified whole. The practice in much of high-energy physics resembles the practice of the past. I would say the same for some medical research. People investigating the role of diet and exercise on depression resemble past researchers, just as physicists do. However, pharma-ceutical research, because of the enormous influence of potential intellectual property rights, is very different from older forms of research. When we discuss the "continuity" versus "break" hypotheses, we need to keep such considera-tions in mind. In the light of current drug research, it seems clear that the champions of the break view of science are right. But with the right sort of political action, we might return to something like the good old days, where the quality of medical research really was very much better.

NOTE

1. See also the FDA's "Drug Approvals and Databases," online at http://www.fda.gov/cder/rdmt/pstable.htm.

REFERENCES

Angell, Marcia. 2004. *The Truth about the Drug Companies: How They Deceive Us and What To Do about It.* New York: Random House.

Antihypertensive and Lipid-Lowering Treatment to Prevent Heart Attack Trial (ALLHAT). 2002. "Major Outcomes in High-Risk Hypertensive Patients Randomized to Angiotensin-Converting Enzyme Inhibitor or Calcium Channel Blocker vs. Diuretic: The Antihypertensive and Lipid-Lowering Treatment to Prevent Heart Attack Trial (ALLHAT)." *Journal of the American Medical Association* 288: 2981–97.

Appel, Lawrence J. 2002. "The Verdict from ALLHAT—Thiazide Diuretics Are the Preferred Initial Therapy for Hypertension." *Journal of the American Medical Association* 288 (23): 3039–42.

Biddle, J. 2007. "Lessons from the Vioxx Debacle: What the Privatization of Science Can Teach Us about Social Epistemology." *Social Epistemology* 21 (1): 21–39.

Black, J. W., and J. S. Stephenson. 1962. "Pharmacology of a New Adrenergic Beta-Receptor Blocking Compound." *Lancet* 2: 311–15.

Brown, James R. 2008. "Community of Science." In *The Challenge of the Social and the Pressure of Practice: Science and Values Revisited*, edited by Martin Carrier, Don Howard, and Janet Kourany, 189–216. Pittsburgh: University of Pittsburgh Press.

———. 2010a. "Medical Market Failures and Their Remedy." In *Science in the Context of Application*, edited by Martin Carrier and Alfred Nordmann. Dordrecht: Springer.

———. 2010b. "One Shot Science." In *The Commodification of Academic Research: Science and the Modern University*, edited by Hans Radder, 90–109. Pittsburgh: University of Pittsburgh Press.

Chasis, H. 1950. "Salt and Protein Restriction: Effects on Blood Pressure." *Journal of the American Medical Association* 142: 711.

Davidson, Richard. 1986. "Sources of Funding and Outcome of Clinical Trials." *Journal of General Internal Medicine* 12 (3): 155–58.

DeAngelis, Catherine. 2006. "The Influence of Money on Medical Science." *Journal of the American Medical Association* (August 23–30).

———, et al. 2004. "Clinical Trial Registration: A Statement from the International Committee of Medical Journal Editors." *Journal of the American Medical Association* (September 15): 1363–64.

Freis, E. D. 1958. "Treatment of Essential Hypertension with Chlorothiazide." *Journal of the American Medical Association* 166: 137–41.

Friedberg, M., B. Saffran, T. Stinson, W. Nelson, and C. Bennett. 1999. "Evaluation of Conflict of Interest in New Drugs Used in Oncology." *Journal of the American Medical Association* 282: 1453–57.

Hamdy, R. C. 2001. "Hypertension: A Turning Point in the History of Medicine . . . and Mankind." Editorial. *Southern Medical Journal* 94 (11): 1045.

Healy, D. 2001. *The Creation of Psychopharmacology*. Cambridge: Harvard University Press.

———. 2003. *Let Them Eat Prozac*. Toronto: Lorimar.

International Committee of Medical Journal Editors. "Uniform Requirements for Manuscripts Submitted to Biomedical Journals: Writing and Editing for Biomedical Publication." Online at http://www.icmje.org/index.html.

Kirsch, Irving , Brett J. Deacon, Tania B. Huedo-Medina, Alan Scoboria, Thomas J. Moore, and Blair T. Johnson. 2008. "Initial Severity and Antidepressant Benefits: A Meta-Analysis of Data Submitted to the Food and Drug Administration." *PLoS Medicine* 5 (2): 260–67. Followed by "Editor's Summary," 268.

Kondro, Wayne, and Barbara Sibbald. 2004. "Drug Company Experts Advised Staff to Withhold Data about SSRI Use in Children." *Canadian Medical Association Journal* 170 (5): 783.

Krimsky, Sheldon. 2003. *Science in the Private Interest*. New York: Rowman & Littlefield.

Lemmens, Trudo. 2004. "Piercing the Veil of Corporate Secrecy about Clinical Trials." *Hastings Center Report* 34 (5): 14–18.

Matthias, A. 2005. "Integrating Research and Development: The Emergence of Rational Drug Design in the Pharmaceutical Industry." *Studies in History and Philosophy of Biological and Biomedical Sciences* 36: 513–37.

Schwartz, W. B. 1949. "The Effect of Sulfanilamide on Salt and Water Excretion in Congestive Heart Failure." *New England Journal of Medicine* 240: 173.

Stelfox, H. T. 1998. "Conflict of Interest in the Debate over Calcium–Channel Antagonists." *New England Journal of Medicine* 338 (2): 101–6.

Whittington, C. J., T. Kendall, P. Fonagy, D. Cottrell, A. Cotgrove, and E. Boddington. 2004. "Selective Serotonin Reuptake Inhibitors in Childhood Depression: Systematic Review of Published versus Unpublished Data." *Lancet* 363: 1341–45.

Willman, D. 2004. "The National Institutes of Health: Public Servant or Private Marketer?" *Los Angeles Times*. December 22, 2004.

Toward a New Culture of Prediction
Computational Modeling in the Era of Desktop Computing

ANN JOHNSON and
JOHANNES LENHARD

COMPUTERS AND SIMULATION METHODS play prominent roles in a wide range of present-day scientific and engineering research. Without doubt, the computer, computational science, and scientific and engineering research have all mutually shaped one another—what is computationally possible informs the questions scientists and engineers ask and the questions scientists and engineers ask, in part, shape the development of new hardware and software. These mutual influences have been frequently examined, largely through the origins of scientific computing and the application of computers to a series of scientific disciplines and problems in the 1940s and 1950s. However, we want to focus on a more recent development, a development that resonates with the break theses this edited volume explores. Namely, we want to argue that the wide and relatively cheap availability of computing power, particularly through mature, networked, desktop computing or the so-called PC (personal computer) revolution, has triggered—but, to be clear, not made inevitable—a reorientation in the practices of scientists and engineers.

Our thesis is that substantial changes have occurred in research practices and culture over the past decade and that these developments have been made possible by everyday accessibility to simulation methods by a wide variety of scientific actors. More precisely, we want to claim that these changes help to generate a highly exploratory mode of research, which, in turn, amplifies the character and role of prediction in science. This new orientation toward prediction challenges some of the basic tenets of the philosophy of science, in which scientific theories and models are predominantly seen as explanatory rather than predictive. The increased value of predictive models leads to what we call a "culture of prediction," common across a wide array of research areas. This culture of prediction, in turn, informs the sort of research projects that scientists and engineers undertake and the kinds of endeavors that funding agencies support.

Mainframe Is Not the Right Frame

Most historical and philosophical accounts of scientific computing concentrate on big science and draw the storyline from the Manhattan Project and the Electronic Numerical Integrator and Computer (ENIAC) to the modern supercomputers. Mainframe computing is portrayed as *the* necessary instrument for the development of computational methods, and the story of computing in science seems to end with the introduction of commercially available mainframe computers in the 1960s.[1] The PC revolution of the 1980s and 1990s is rarely mentioned at all; when personal computers appear, they appear as the smaller, unimportant, even trivial siblings of bigger computing machines. According to the orthodox account, mainframes display all the interesting features and their development is more accessible to study. The PC is often portrayed as commercially important but philosophically meaningless. We do not agree with this view.

Furthermore, we hold that it is precisely desktop computing that has changed scientific practices and capacities over the past two decades.[2] Computing here is not restricted to hardware, but includes software and all elements of computational modeling. Although IBM introduced the personal computer in 1981, it was not until the 1990s that personal computers offered the computing power, networked connectivity, and software packages necessary to support scientific enterprises in substantial ways. By the 1990s desktop computing was maturing, instantiated through workstations, such as those from Sun and Silicon Graphics, or clusters of Intel-powered machines, dependent on Ethernet and Internet connections as well as the continual miniaturization of integrated circuits, and therefore the increased power and decreased cost of general computing, characterized by Moore's Law.

It is uncontroversial that a broad range of computer simulation techniques is now running on desktop computers. The first steps and ingenious ideas taken by Stanislaw Ulam, John von Neumann, and Nick Metropolis, for instance, to implement a Monte Carlo method on computing machines took place in the pioneering phase where computers were "one-off" prototypes constructed for specific uses and as experiments in computing themselves. In contrast, today's networked PCs provide an environment that supports multiple, commercially available, random-number generators and Monte Carlo simulation modules for users who need not be programmers, suitable for use on a variety of different machines. Practically all simulation techniques have become widespread and easily available in commercial or open-source software packages. Finite element methods provide another prominent example. Finite element software has become a multimillion-dollar business, with nearly as much money being wielded in advertisement as in software development. These packages run on desktop computers affordable to individual, noncommercial users and the smallest scientific operations—in effect going a long way to level the field between the biggest university computing centers and individual scientists and engineers.

Consequently, in looking at actual computational practices in recent science, we argue that wide availability matters. Of course, this is correct in a trivial and circular sense, as any simplistic measure of the influence of an instrument on scientific practice and culture depends on wide availability and use. There are, however, nontrivial, even surprising, effects of the adoption of the PC on the organization, social relations, and the epistemology of science. To give a full account of these changes would require an extensive historical-philosophical research project. While we have undertaken such a project, this chapter provides a brief sketch of our claim and why it is justified.

Mathematization and Prediction

The connection between science and mathematics is a rich and deep issue with its own lengthy history. Efforts to produce mathematical models of motion and the heavens surely constitute a chief characteristic of the scientific revolution—arguments about the existence of such an event notwithstanding. Although a scientific revolution per se requires many dimensions—epistemological, social, political, economic, and so on—in which revolutions occur, it is widely accepted that, following E. J. Dijksterhuis (1986), the "mechanization of the world picture" was a radical redefinition of what it meant to know. Knowing came to mean being able to represent a phenomenon as a mathematical model. Mathematical knowledge was especially powerful because it facilitated prediction. In astronomy this was clear; it was useful to know when various phenomena were going to occur for a variety of purposes—principally among them religious observances, time keeping, and navigation. However, the predictive claims made by scientists and engineers clearly depended on mathematization; mathematical models alone facilitated prediction. And while mathematization changed what it meant to know, it is an even stronger claim to argue that mathematically determined predictions were precisely the element of mathematization that redefined what it meant to know. Knowing a system came to mean knowing what would happen next—it meant predicting unknown, even surprising, events and objects; when such events and objects were experienced, then the model must be true and its further predictions taken seriously.

Since the scientific revolution, prediction has been one of the dominant focuses of scientific activity—and an area in which the sciences claim a special facility. Predictions are epistemologically significant because they are clearly verifiable (unlike explanations); many can be shown to be accurate or not. Predictions are therefore preconditions for explanations. In other words an explanation cannot explain anything if it is not verifiable; verification often means checking the predictions of a model against real-world events. In this sense philosophers have often assigned prediction a secondary role to explanation, which has long been seen as the goal of scientific theories. Carl G. Hempel (1965), with the deductive-nomological model of explanation, provides an example by seeing explanation and prediction as two sides of one coin. However, the still ongoing extensive discussion in the philosophy of science concentrates on explanation as the key issue and proposes a unified view that takes explanation as embedding results into a wider framework, or a causal account of explanation that centers on finding causal chains. We do not claim that prediction replaces explanation as a function of scientific models, but rather that it warrants attention independent of its role in explanation.

The "no miracles" argument for realism provides another instance where the philosophy of science refers to prediction. This argument states that scientific models of nature must be true if their predictions are true, since it would be a miracle if random formulae produced an accurate prediction. Again, prediction plays the role of handmaiden to another notion, namely realism. Problematizing prediction itself and decoupling it from shopworn debates about realism or scientific explanation frames our examination of the role of ubiquitous computing in present-day scientific and engineering research.

Given this background, when electronic computers became available to scientists and engineers in the mid-twentieth century, computational models became more attractive and gradually the nature of prediction changed. Computational models allowed far more variables to be incorporated and for mathematically intractable models to be remodeled as numerically approximative models. As a result, far more physical systems could be modeled, but ontological problems also emerged. Principally, scientists asked whether similarities between computational models and physical systems were structurally meaningful; if a computational model yielded a result that indeed became verified, did this mean it accurately captured, in mathematical terms, the mechanism of the system? Clearly this implication was not true for many computational models.

Yet in the earlier scientific revolution account of mathematical models, ontological resemblance was significant. Many natural philosophers claimed that their mathematical models were *true* representations of the mechanics of the physical system. Models had two possible verification claims—whether their predictive output was, in fact, accurate and whether a model was believed to be a true mathematical representation of nature. However, in the present-day case, no one is claiming that computational models involve this kind of ontological similarity; in fact, simulations are dependent on computational mechanisms that unarguably have no counterpart in reality—like Monte Carlo simulations or fictional and arbitrary finite elements. Clearly, most simulationists do not claim that nature works in an iterative approximation mode. Computational models' chief attraction is that they produce a predictive answer, not a mimetic model of the causal mechanism of a real-world phenomenon. On the basis of predictive success, reasoning about a model's virtues to be a "good enough" representation is not ruled out by our viewpoint, indeed this often takes place.

Opacity and Complexity of Computational Modeling

We tend to view computational models as a special kind of mathematical model —a kind that does not dissolve but rather amplifies the problem of ontological resemblance. One result of discarding the need for a tight connection between a physical phenomenon and a computation model of it is that all sorts of phenomena can be modeled, even when the mechanisms are poorly understood. As an example, consider the controversy between weak and strong approaches to artificial intelligence (AI). Strong AI is concerned with modeling human intelligence by implementing similar mechanisms into computational machines. On the other hand, the weak approach imitates behavior regardless of the similarity of mechanisms.

The controversial debate in AI is exemplified by Alan Turing's famous "weak" proposal of how to test whether a machine is intelligent and a great number of follow-up contributions that debate weak versus strong claims (Turing 1950). The weak program predicts but may not allow any insight into how actual human intelligence functions; the strong program holds that the point of AI is to model the processes of intelligence. This example is instructive in showing that true similarity to a real-world system may need to be traded off for the production of predictive models; the complexity of physical systems is often too great to allow mathematical models of the same scale as physical systems. Simplification, idealization, and approximations are needed to make predictive models, but they come at the cost of fidelity to real physical systems. This is true of many different classes of simulation models, and scientific practitioners must sacrifice either fidelity or predictive output.

In point of fact, most simulationists choose to build and use predictive models—some practitioners call these "theory-light" models. Representational conformity to the physical system gets trumped to make models of complex systems. Consequently, the computer becomes an instrument for dealing with complexity. But ironically, the computer is itself complex, so while trying to produce predictive accounts, it is itself opaque.[3] This shows a distinction between traditional mathematical models and the special case of computational mathematical models. Stating it bluntly, simulation models do not determine model behavior by model structure, because a number of unassigned parameters and black-boxed elements are normally included in the model's structure that first have to be adapted before any model behavior can be tested or compared with any expectations of scientists.

In this process monitoring provisional output and optimizing the overall fit is an essential part of the modeling process. While earlier mathematical models were aiming to produce a transparent causal explanation, computational models are opaque, what happens inside the model is not clear and in some cases the reason for the coherence of the model and reality isn't explainable. This is a surprising new phenomenon, because mathematics is usually expected to introduce transparency. The failed expectation of transparency extends to the computer itself. Historian Michael S. Mahoney (2011) claims that while the computer is accessible for mathematics in principle, in practice it is not. He is right; viewed step-by-step, even the most complex computer models satisfy the strictest formal standards, but viewed as a whole—that is, dealing with the model performance that results out of complicated interactions—things become opaque. On the basis of this argumentation, performance is not so much the endpoint of the modeling process, but rather the starting point. Epistemic opacity and the lack of ontological referents emerge from the layering of models and the distance from the real world. As opposed to the no-miracles argument, which claims that a match between the output of a mathematical model and experimental data would be miraculous if there were not some truth to the model, there can be a plurality of different models that align with experimental data since there are multiple routes to coherence with empirical data or the real world. Distinctions between different models must be made through tinkering rather than through a single test of prediction; computational complexity prevents assessment being based on representational conformity.

Consider artificial neural networks (ANNs) as an example of model performance acting as a starting point for research. ANNs were initially proposed because of their partial similarity to physiological neurons. However, their application to more widespread problems depended on a so-called learning algorithm (backward propagation). Learning algorithms allowed the adaptation of input-output behavior through iterative adjustments of a large number of model parameters. The specific dynamical properties of such a network hinged on the special parameter value assignments, whereas the structure remained invariant. This meant that the construction of the model was only the starting point for research with artificial neural networks. The vast number of parameters and their interactions generated the model's complexity, and they focused researchers' attention on the specification of parameters rather than the structure of the model itself.

Simulations, the PC, and an Exploratory Mode of Research

Thus computational models are opaque, and work on them has to proceed by experimentation. Researchers experiment on simulations by changing, adding, and adapting parameterization schemes and submodules and running the simulations repeatedly. Such a practice rules out expensive and time-consuming computation techniques for most research projects; models need to be processed within hours, not months. In an era when computing meant waiting for time on a mainframe CPU, "tweaking" models—that is, running iterative versions of models with minor changes—was impractical. The exploratory nature of computational modeling is therefore dependent on cheap, convenient, and local access to computational power. This accessibility comes only with the mature, networked desktop computer. Historically, this means the widespread change in modeling practice occurred in the 1990s—not in principle with the introduction of computers into scientific practice in the 1940s and 1950s, nor with widespread use of mainframes, which were neither cheap nor convenient, in the 1960s and 1970s.

Computational research became better integrated to both theory and experiment and evolved into an indispensable dimension of the scientific enterprise, as opposed to the kind of computation that was common in the 1950s through 1980s, which was a more independent branch of investigation. Now researchers work on computational models, whether they consider themselves experimentalists or theorists. "Working on" models means exploring the relationships between input data and output data to produce either predictions or better models. With the mature desktop computer, model exploration becomes a key practice in scientific research, and not one relegated to computational scientists, but rather one that is commonplace among most scientists and engineers.

This exploratory mode depends on researchers' ability to quickly assess the outputs of various models. Assessment of model outputs is especially feasible in visual, as opposed to numerical, forms. Visualization is also a feature of the desktop computing revolution, albeit one that did not emerge from scientific computing. Desktop computers, with their many different kinds of users, have been at the center of a series of changes in the visual display of information. From computer games to animation to web pages, computers since the 1980s have grown to be devices focused on visual display. These extrascientific demands have created capacities critical to scientific uses—especially in three-dimensional display. Visualization in science has changed because of the ability of computers to generate images. These images are tremendously powerful; they carry information more efficiently than do tables of numerical outputs, as a result they yield compelling results—sometimes in misleading ways. Consequently, visualization reinforces the exploratory mode of scientific research, by making possible the quick uptake of results from computational models. While there is not a direct causal link between exploration and visualization, both depend on fast, cheap, accessible computing, which only emerges in the era of desktop computing. This amounts to a change in scientific culture: scientific research proceeds by experimenting on computational models and assessing visual outputs, replacing efforts to write the one computational model that will yield all the right results. The new strategy is exploratory, evolutionary, adaptive, and provisional.

This exploratory mode of scientific investigation depends on the development of a research technology, using that phrase in Terry Shinn's understanding. Shinn (2001) claims that four elements characterize research technologies. First, they are produced by interstitial communities; that is, they do not arise from single institutional, disciplinary, or industrial problems or uses. Second, "the devices that research-technologists deal with are generic" (Shinn 2001, 9), meaning they are not designed to respond to any specific industrial or academic demand. Third, research technologies can be seen as carriers of new metrologies, that is "they generate novel ways of representing visually or otherwise events and empirical phenomena." Lastly, they are disembedded from their context of invention, a direct consequence of their genericity, they become nonlocal to any one scientific community. The software packages necessary to produce simulation models for desktop computers seem to share these characteristics, although they are clearly not instruments in the traditional sense. Thus we claim that the software packages themselves are a new kind of research technologies and benefit from being conceived as such. This status allows them to play a key role in generating a new investigative mode in scientific and technological research.

Software packages to perform finite element analysis, continuous fluid dynamics, Monte Carlo simulations, and others too numerous to mention are a big business. Dozens of companies around the world make millions if not billions of dollars selling such software to scientific and engineering researchers. Thus this software is a commodity, which further reinforces the claim that it acts as a research-technology. Commodification implies several of Shinn's elements—these packages operate across disciplines, across the public-private divide, they are disembedded from the context of their invention in order to be widely sold, and they must be generic to have such a broad base of users. Black-boxed software packages are commercial goods that depend, in part, on their hiding of structures and details implemented in the software—the practitioner does not have nor need knowledge of them. Computational modeling has itself become commodified. The commercial availability of such software enables the exploratory mode of scientific investigation since researchers can buy the modules rather than spending their time writing the programs themselves. The movement of the packages plus the local adaptation of software creates a networked social dimension to scientific culture; the common use of software facilitates the exchange of information and the growth of scientific knowledge between research teams in different locations. Hence the social networking aspect constitutes a decisive intersection where institutional organization and research technology interact.

Science and the Culture of Prediction

Thus it is our claim that the personal computer has generated a substantial change in scientific practice. But it can only do so through the use of modular software packages that have become commercially available over the past generation. The commercial nature of this software is only possible given the wide availability of a small number of hardware platforms. The existence of a system of hardware and software spanning disciplinary lines and industrial-academic divides facilitates an exploratory mode of working with simulations. This system was not in evidence in earlier versions of simulation practice, prior to the maturation of the desktop computer. Thus this element of scientific practice only emerges in the 1990s. This change is important in itself, but also because it marks a wider turn toward valuing prediction. These simulation practices focus on producing predictions of system behavior, some of which are experimentally verifiable, some of which are not experimentally accessible. The verification of some predictions lends credibility to many more that are not directly verifiable.

While this attitude is not logically justified, it does illustrate the cultural traction of predictive models. Conversely, because there is a culture of prediction, predictive models are valued even in the absence of verification, often because they share simulation methods with verified models. We acknowledge that prediction has a long history in the mathematical sciences; however, the role of prediction has changed with the desktop computing revolution. The wide availability of hardware and commercial simulation software since the 1990s has facilitated prediction to acquire a more prominent role. Due to the importance of prediction, there emerged—and still is in a formational phase—a full-fledged "culture of prediction." This claim includes both the specificity and impact: the formation of this culture is recent and has had a wide impact both within and outside of the scientific enterprise.

We did not invoke bold claims regarding the break thesis so prominently addressed in this edited volume. However, quite obviously our chapter has relations to this thesis and in particular to the ways it is discussed by other contributions in this volume. Alfred Nordmann made a case for technoscience and its main or even sole aim to achieve control. Predictive capacity is in fact key to control, and in a sense we advocate scientific instrumentation for control. We dare to join the technoscience-control discussion because we think that the culture of prediction can be investigated separately so that we can avoid dealing with the controversial terminological load of technoscience.

Martin Carrier, on the other side (of the break), has argued in his chapter in this volume against seemingly hasty ways to establish technoscience because any approach solely directed to control would fail to achieve a certain degree of sustainability. This would hinge on a more profound theoretical understanding. He foresees a quick and inglorious ending for these approaches. We would like to add to his discussion that the provisional nature of computational models has been identified as an explicit part of the culture of prediction. It is the speed, ease, and versatility of computational modeling that potentially compensate for the fact that any single model may have a relatively limited lifespan. This would even reinforce our claim for a newly emerging culture of prediction, a culture that—openly or implicitly—accepts missing sustainability. Again, this view recognizes parallels and possible interrelations to broader streams of nonscientific culture.

NOTES

1. An exception is November 2011.
2. By "scientific practice" we follow the meaning established by Andrew Pickering in his *The Mangle of Practice*—that is, what scientists do, "the acts of making and unmaking that [scientists] perform in the field" (Pickering 1995, 3). Pickering also defines scientific culture in a way useful to us—"the field of resources that practice operates in and on." Scientific practice and culture cannot be said to change with the genesis of new techniques or protocols, but rather with their widespread dissemination. In the case of simulation, dissemination is dependent on cheaper, smaller, more powerful computing technology; therefore, we focus on the PC revolution when seeking changes in practice.
3. Perhaps then we do have some similarity between the physical processes that are being modeled and the models themselves. However, if epistemic opacity is what is similar, then this isn't a desirable kind of similarity.

REFERENCES

Dijksterhuis, Eduard J. 1986. *Mechanization of the World Picture: Pythagoras to Newton.* Princeton, N.J.: Princeton University Press.

Hempel, Carl G. 1965. *Aspects of Scientific Explanation.* New York: Free Press.

Mahoney, Michael S. 2011. *Histories of Computing.* Cambridge: Harvard University Press.

November, Joseph. 2011. *Digitizing Life: The Rise of Biomedical Computing in the United States.* Baltimore, Md.: Johns Hopkins University Press.

Pickering, Andrew. 1995. *The Mangle of Practice.* Chicago: University of Chicago Press.

Shinn, Terry. 2001. "A Fresh Look at Instrumentation: An Introduction." In *Instrumentation between Science, State, and Industry,* edited by B. Joerges and Terry Shinn, 1–13. Dordrecht: Kluwer Academic Publishers.

Turing, Alan. 1950. "Computing Machinery and Intelligence." *Mind* (236): 433–60.

Epilogue
The Sticking Points of the Epochal Break Thesis

HANS RADDER

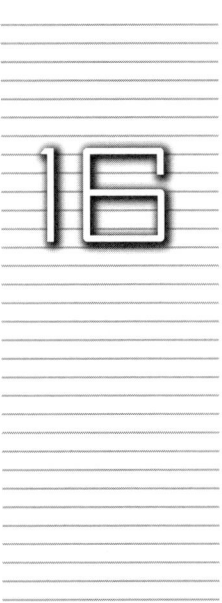

AT THIS POINT IT SHOULD BE CLEAR that the epochal break thesis involves a wide-ranging and ambitious claim concerning recent science and its history. The preceding chapters display a diversity of views on this thesis. Hence, drawing a single, straightforward conclusion from these chapters, for or against the epochal break thesis, would obviously be premature. What is feasible, however, is to extract from the preceding chapters a number of "sticking points," to use a phrase of Ian Hacking; that is to say, central issues that can be expected to remain the focus of extensive research and debate. Each of these issues entails a variety of further research questions. To be sure, not all of these issues are completely new. Yet some are relatively new within certain disciplines, such as the study of the science-technology relationship within mainstream philosophy of science or the exploration of the commercialization of science within science and technology studies.

Moreover, although the chapters collected in this edited volume provide an important starting point, in view of the high stakes of the epochal break thesis all of these issues deserve more detailed and sustained analyses and assessments. The sticking points briefly described below include historical and theoretical accounts of the role of old and new methodologies, demands for empirical and conceptual clarification of the central notions at stake in the debate, ontological and epistemological issues concerning the nature and development of the sciences, social-scientific inquiry into the relationships between science, technology and the wider society, normative concerns about the sociocultural position of recent science and technology, and historiographical questions of how to support or challenge a comprehensive historical thesis like the epochal break thesis.

1.

Rightly or wrongly, the traditional view of science was often based on the example of theoretical physics. As the contributions to this book show, the epochal break thesis urges us to focus on other disciplines, such as climate science, environmental science, biomedical science, or computational science. Furthermore, it encourages the examination of possible and actual interactions between such disciplines. From this perspective different modes of inquiry have to be addressed: not exclusively theoretical representation or explanation but rather experimental intervention, data processing, visualization, modeling, computer simulation, and prediction. Finally, since the knowledge in question is supposed to be applied, or even developed, within the wider society, the issue of the cultural meaning and the social acceptance or rejection of this knowledge arises. For this reason, in addition to these disciplines, the social sciences and humanities are involved and need to be taken into account. One may argue that scientific expert knowledge also has to be complemented by the lay knowledge of the people who may be affected by the realization of the technoscientific or mode-2 projects. This, in turn, requires explicit methodological reflection on the relationship between expert and lay knowledge and on the question of how lay knowledge may be taken into account. In sum, an important question is whether technoscientific or mode-2 projects require a distinct methodology that differs from (what is often taken to be) the methodology of theoretical physics.

2.

Whatever one may think of the epochal break thesis itself, the notions employed in its different articulations—primarily, technoscience, mode-1/mode-2 science, and postacademic and postnormal science—are important and hence deserve further inquiry. As indicated in the introductory chapter, these labels are often introduced and used in rather loose ways. Consider the notion of technoscience, which has been the focus of quite a few of the preceding chapters.

Although some accounts of this notion are available, none is very detailed. Hence, what we need is a well-developed and plausible explication of the notion of technoscience. Furthermore, how does this notion relate to "science," to "technology," to a "technological style," to "engineering science," to "applied science," to "industrial science," to "entrepreneurial science," and the like? Analogous questions may be posed concerning the notions of mode-1 and mode-2 and the concepts of postacademic and postnormal science. To be sure, such explications should not be limited to conceptual analysis but also employ the results of empirical studies. In this way we may examine, for instance, whether technoscience may coexist with academic science or mode-1 with mode-2 science, and the extent to which such coexistence occurs in actual scientific projects, research programs, or academic disciplines.

3.

If true, the epochal break thesis entails significant ontological and epistemological questions. If science has been transformed into technoscience, does this imply that, in contrast to the objects of earlier natural sciences, the objects of current "natural" sciences are artificial, humanly contrived entities? If so, this clearly goes against realist ontologies, which assume the existence of a human-independent nature. Hence, this ontological antirealism would be a consequence of the contingent development of science itself rather than a position that could be vindicated, or criticized, by philosophical argument. A further ontological issue arises from the technoscience of human beings. For instance, what are the ontological implications of seeing a neural network as a model of the human mind or a robot as a model of a human being? Epistemologically, there is the question whether, and if so to what extent, the new modes of inquiry employ different epistemic criteria. If so, do these criteria have the same epistemic force? For instance, do computer simulations or colorful visualizations have the same probative force as quantitative experimental tests or theoretical inferences? How should we compare the epistemic value of knowing how something works with understanding what it is? And how should the idea of "social robustness," put forward by mode-2 theorists as a novel epistemological criterion, be defined and assessed? Does it replace, or complement, the mode-1 criterion of peer review? Finally, there is the notion of uncertainty, which is taken to be a core notion of postnormal science. Should we seek the roots of uncertainty at an ontological, an epistemological, or a social level—or perhaps at all three levels simultaneously?

4.

As has been shown in detail in the sociology of science, scientific practices and scientific knowledge have never been isolated from the wider society. Since its inception, science has always possessed both a sociocultural significance and a technological potential. Yet, exploring the epochal break thesis involves examining the relationship between science and society from a more specific perspective, with a focus on the search for novel patterns in that relationship. A major subject is the connection between science and technology, which may be studied from an empirical, a conceptual, or an evaluative point of view. Related to this is the strongly increased commercialization (or broader, commodification) of academic science. It is also important to realize that even the mere proclamation of the occurrence of an epochal break may bring about substantial social effects, and hence we should take into account the social causes and impacts of the emergence of epochal break rhetoric. In addition, many specific developments carry an immediate social significance. Just think of the many social issues surrounding the uses of biomedical science, the risks involved in realizing chemical or physical technologies, the dangers of the continuing expansion of science-based war technologies, the privacy issues raised by the large-scale employment of automated processes, or, rather differently, the uses of technoscience for purposes of entertainment.

5.

The overarching normative question concerns the social and moral desirability and legitimacy of an (alleged) epochal break. Investigating this question includes a critical analysis of unequal power relations in technoscientific, mode-2, or postacademic and postnormal practices. Because of the dominance of a neoliberal worldview and a neoliberal politics during the past decades, the scholarly study of science has virtually ignored the issue of power. More recently, however, there seems to be a renewed interest in this issue, in particular in studies of the commercialization of academic science. Related to this is the normative debate about the future of the university. Should we continue to transform our universities into hierarchical, corporate organizations, fostering an entrepreneurial ethos with its strong emphasis on grantsmanship, capitalization of knowledge, and public relations? Or should we seek to recover an academic ethos, according to which the social legitimacy of science does not derive from its short-term exploitation for private gain, but from its provision of a knowledge infrastructure that constitutes a long-term resource for the public good? In addition to these fundamental normative questions concerning the aims of science and our scientific institutions, the developments mentioned under the preceding point obviously entail a host of more specific normative issues.

6.

Finally, as a grand historical and philosophical narrative, the epochal break thesis requires an in-depth discussion of historical methodology, or even broader, of our philosophical conception of history and historiography. Which approach to the issue of how we make sense of history is implied by the thesis? How might such a comprehensive thesis be plausibly supported by the results of particular historical case studies? During the past decades grand narratives have been severely criticized from various perspectives. Just think of the critiques of universal theories of scientific rationality, the rejection of the theory of technological determinism, or the decline of Marxism and other critical social theories. Yet, taking the epochal break thesis seriously requires that we reexamine such critiques and bracket any immediate rejection of the possible existence of broad historical patterns. At issue is the question of historical periodization, including the identification of periods and the nature of the transition processes from one period to another. For instance, present-day historians often claim that there was no such thing as the Scientific Revolution in the sixteenth and seventeenth century. In this case we should ask which historical methodology underlies this claim, which alternative methodologies would license the occurrence of a Scientific Revolution, and which are the arguments for or against these methodologies? In this way, debating claims of an epochal break leads us to reflect on fundamental questions concerning the nature and purpose of historical reasoning.

Mieke Boon is an associate professor at the philosophy department of the University of Twente. She holds MSc and PhD degrees in the engineering sciences. After conducting laboratory research for fifteen years in biotechnology for the mining industry and environmental technologies, she moved to her current research topic of developing a philosophy of science for the engineering sciences. For this she was awarded the Vidi grant of the National Dutch Science foundation. Boon has published on these themes in *International Studies in the Philosophy of Science, European Journal of the Philosophy of Science, Metascience, Perspectives on Science,* and *Techné,* as well as in several edited books.

James Robert Brown is a professor of philosophy at the University of Toronto. His interests include a wide range of topics in the philosophy of science and mathematics: thought experiments, foundational issues in mathematics and physics, visual reasoning, and issues involving science and society, such as the role of commercialization in medical research. His books include *The Rational and the Social; The Laboratory of the Mind: Thought Experiments in the Natural Science; Smoke and Mirrors: How Science Reflects Reality; Philosophy of Mathematics: An Introduction to the World of Proofs and Pictures; Who Rules in Science: An Opinionated Guide to the Wars;* and *Platonism, Naturalism and Mathematical Knowledge.*

Martin Carrier is a professor of philosophy at Bielefeld University and part of the Institute of Science and Technology Studies. His chief area of work is the philosophy of science—in particular, historical changes in science and scientific method, theory-ladenness and empirical testability, intertheoretic relations and reductionism, and methodological issues of application-oriented research.

Carrier was awarded the Leibniz Prize of the German Research Association for 2008. His recent books include *Wissenschaftstheorie: Zur Einführung* (Introduction to the Philosophy of Science) and *Raum-Zeit* (Space-time). His recent edited volumes include *The Challenge of the Social and the Pressure of Practice: Science and Values Revisited* (with Don Howard and Janet Kourany) and *Science in the Context of Application: Methodological Change, Conceptual Transformation, Cultural Reorientation* (with Alfred Nordmann).

Valerie L. Hanson is an associate professor of writing at Philadelphia University. She received her PhD from Pennsylvania State University in English, rhetoric, and composition; her dissertation focused on the rhetorics of nanotechnology. She writes about ethics and technology as well as the intersections of science, rhetoric, and technology, such as the impact of visualization technologies on the formation of scientific fields like nanotechnology. Recent articles have appeared (or are forthcoming) in *Science as Culture* and *Science Communication*; additional forthcoming work includes a book about the rhetorics of digital images in nanotechnology.

Andrew Jamison is a professor of technology, environment, and society at the Department of Development and Planning at Aalborg University. He has a BA in history and science from Harvard University and a PhD in theory of science from Gothenburg University. Jamison has carried out research on social movements, science and technology policy, and environmental politics and is the author, most recently, of *The Making of Green Knowledge: Environmental Politics and Cultural Transformation*; *Hubris and Hybrids: A Cultural History of Science and Technology* (with Mikael Hård); and *A Hybrid Imagination: Science and Technology in Cultural Perspective* (with Steen Hyldgaard Christensen and Lars Botin).

Ann Johnson is an associate professor of history at the University of South Carolina. She is the author of *Hitting the Brakes: Engineering Design and the Production of Knowledge* and is currently working on a book with Johannes Lenhard about mathematization, computational models and simulations, and the role of information technologies in the changing epistemologies and cultures of science and engineering in the dawn of the twenty-first century.

Tarja Knuuttila is a senior research associate in philosophy at the University of Helsinki. She holds degrees in economics and business administration (MSc from the Helsinki School of Economics) and in philosophy (MA and PhD from the University of Helsinki). The main themes of her work have been modeling and scientific representation, the methodology of economics, as well as the commodification of science. Knuuttila has published on these themes in *Biology and Philosophy*; *Erkenntnis*; *Forum: Qualitative Social Research*; *Philosophy of Science*; *Semiotica*; *Studies in History and Philosophy of Science*; *Science Studies*; *Science, Technology, and Human Values*; and *Sociology of the Sciences Yearbook*; as well as in numerous edited books.

Angela Krewani is a professor of media studies at Marburg University. She is the author of *Moderne und Weiblichkeit: Amerikanische Schriftstellerinnen in Paris* and *Hybrid Forms: New British Cinema–Television Drama–Hypermedia*. Krewani edited *Artefacts/Artefictions: Transformational Processes in Contemporary Literatures, Media, Arts, and Architectures*, and coedited *Hollywood: Recent Developments*. She is currently focusing on the hybridization of media systems and new forms of media narration. She has also published on the image in such contemporary natural sciences as biomedicine and nanotechnology. From 2006 through 2007 she was a fellow at the Center for Interdisciplinary Studies at the University of Bielefeld. In 2008 she was a visiting professor at Brooklyn College in New York.

Wolfgang Krohn is professor emeritus for science and technology studies at the University of Bielefeld. His main research interests encompass the origin of modern science and the philosophy of Francis Bacon, transdisciplinary research and real-world experiments, and the aesthetics of science and its relation to research in the arts. A selection of recent publications includes *Methoden transdisziplinärer Forschung* (Methods of Transdisciplinary Research), with Matthias Bergmann and others; *Nachrichten aus der Wissensgesellschaft: Analysen zur Veränderung der Wissenschaft* (Messages from the Knowledge Society), with Peter Weingart and Martin Carrier; and a contribution to the *Oxford Handbook of Interdisciplinarity*.

Chunglin Kwa is a lecturer in science and technology studies at the University of Amsterdam. He teaches courses in the philosophy of science and the history of science and on policy aspects of global climate change. On the latter subject he has published in *Social Studies of Science* and *Science & Public Policy*. His other research interests include the history of representations of the landscape. Articles on this subject have appeared in *The European Legacy* and *Configura-*

tions. The English translation of Kwa's book *Styles of Knowing: A New History of Science from Ancient Times to the Present* was recently published by the University of Pittsburgh Press.

Johannes Lenhard studied philosophy and mathematics with a dissertation in mathematics at the universities of Heidelberg and Frankfurt. He has held positions as research assistant at Bielefeld's Center for Interdisciplinary Research and as research associate professor at the University of South Carolina. Currently, Lenhard is affiliated with the philosophy department at the University of Bielefeld. His main area of research is philosophy of science, and he is investigating the various ways in which computers as new instruments influence science.

Cyrus C. M. Mody is an assistant professor in the history department at Rice University. His research focuses on university-industry-government interactions in the field of microelectronics research between 1965 and 2005—the era of the putative epochal break. His forthcoming book, *Instrumental Community: Probe Microscopy and the Path to Nanotechnology*, examines the invention, spread, and commercialization of the scanning tunneling microscope and atomic force microscope—two of the iconic instruments of the post-1980 era of declining corporate basic research and increasingly commercialized academic research.

Alfred Nordmann received his PhD in Hamburg and then served on the faculty of the philosophy department at the University of South Carolina, where he continues to be affiliated. In 2002 he became a professor of philosophy and history of science at Darmstadt Technical University. He has investigated conceptions of science in the work of Georg Christoph Lichtenberg, Charles Darwin, Heinrich Hertz, and Ludwig Wittgenstein. Nordmann is currently focusing on the development of a comprehensive philosophy of technoscience. As rapporteur for a European expert group, he produced the 2004 report on *Converging Technologies for European Knowledge Societies*. He coedited several volumes on the philosophical dimensions of nanotechnological research and most recently, with Martin Carrier, *Science in the Context of Application.*

Hans Radder is a professor of philosophy of science and technology at VU University in Amsterdam. Principal themes in his work are scientific observation and experimentation; the historical, epistemological, and ontological nature and role of concepts; the issue of scientific realism; and the normative and political significance of science and technology. Radder is the author of *The World Observed/The World Conceived*; *In and about the World: Philosophical Studies of Science and Technology*; and *The Material Realization of Science*. He is also the editor of *The Philosophy of Scientific Experimentation* and *The Commodification of Academic Research: Science and the Modern University*.

Gregor Schiemann is a professor of history and philosophy of science at the Department of Philosophy at the Bergische University in Wuppertal. He holds a diploma in physics and a PhD in philosophy. His areas of specialization include the history and philosophy of physics in the nineteenth and twentieth centuries and the concept of nature. He is coeditor of the *Journal for General Philosophy of Science*. Recent monographs include *Hermann von Helmholtz's Mechanism: The Loss of Certainty* and *Werner Heisenberg*. With Michael Heidelberger, Schiemann edited *The Significance of the Hypothetical in the Natural Sciences*.

Astrid Schwarz was trained in Germany and France in philosophy and biology. She is currently based at the Department of Philosophy at Darmstadt Technical University and at the Program for Science Studies at the University of Basel. Her field is the philosophy and cultural history of science and technology. Case studies are drawn from ecology, the environmental sciences, nanotechnology. Schwarz is coeditor of the *Handbook of Ecological Concepts* (the first volume, *Ecology Revisited: Reflecting on Concepts, Advancing Science*, appeared in 2011). With Angela Krewani, she initiated a Web-based information system for research and education, *Visual Cultures of Ecological Research*. She is currently engaged in the German-French project Genesis and Ontology of Technoscientific Objects. These interests coalesce in her forthcoming book, *De-Liberating the Experiment: Trial and Error beyond the Laboratory*.

Jutta Weber is a philosopher, media theorist, and science and technology studies scholar. She is the interim professor for media, culture, and society at the University of Paderborn in Germany. Her research focuses on epistemological, ontological, and sociopolitical dimensions of technoscientific knowledge production and the interdependencies of the technosciences and everyday life. Weber held visiting professorships in several countries, participated in an EU project on artificial intelligence, robotics, and ethics and is a member of the EU Cost Action Living in Surveillance Societies (LISS).

Note: Page numbers in italic type indicate photographs.